DATE DUE FOR RETURN

This book may be recalled before the above date.

Fundamentals of
Supercritical Fluids

Fundamentals of
Supercritical Fluids

Tony Clifford

Professor of Chemical Technology
University of Leeds, Leeds, UK

This book has been printed digitally in order to ensure its continuing availability

OXFORD
UNIVERSITY PRESS

Great Clarendon Street, Oxford OX2 6DP

Oxford University Press is a department of the University of Oxford.
It furthers the University's objective of excellence in research, scholarship,
and education by publishing worldwide in

Oxford New York

Auckland Bangkok Buenos Aires Cape Town Chennai
Dar es Salaam Delhi Hong Kong Istanbul Karachi Kolkata
Kuala Lumpur Madrid Melbourne Mexico City Mumbai Nairobi
São Paulo Shanghai Singapore Taipei Tokyo Toronto

with an associated company in Berlin

Oxford is a registered trade mark of Oxford University Press
in the UK and in certain other countries

Published in the United States
by Oxford University Press Inc., New York

© Oxford University Press, 1999

A catalogue record for this book is available from the British Library

Library of Congress Cataloging in Publication Data
Fundamentals of supercritical fluids / Tony Clifford.
1. Supercritical fluids. I. Title.
TP156.E8C55 1998 543'.0896—dc21 98-19460

ISBN 0-19-850137-4 (Hbk)

Preface

This book is intended as an introduction to the principles governing the use of supercritical fluids. As such, it is not a review of previous work, and published experimental work is included only to illustrate these principles. Because of the easy access to data and figures, much of the work quoted is from our own research group, although this represents only a small fraction of published work in this area. Because the book is intended for explanation rather than review, references are given only to major developments, or when it may be useful for readers to follow up a subject further, for example, to obtain experimental details, which are only given briefly here. The first four chapters are on the basic aspects of thermodynamic properties and the second four chapters on particular areas of study, mainly separation science. There are other aspects of work in supercritical fluids not covered, notably particle formation, surface chemistry, spectroscopy, and supercritical water oxidation, although the first four chapters have information in them relevant to these topics.

I have gained my knowledge of the subject, such as it is, by working with a large number of people, to whom I am very grateful. In particular, I mention my colleague Keith Bartle, with whom I began working in the area. More recently, Chris Rayner has joined those at the University of Leeds working in supercritical fluids. The more difficult modelling work described in the book has been carried out by a colleague Zhu Shuang; much of this is not published elsewhere. There are also many colleagues outside the University, both in the UK and abroad, with whom I have had fruitful relationships. Prominent amongst these is Steve Hawthorne at the University of North Dakota, also Bill Wakeham of Imperial College, Andy Johns of the National Engineering Laboratory, Neil Smart of British Nuclear Fuels, Chien Wai of the University of Moscow, Idaho, and Zenko Yoshida of the Japanese Atomic Energy Research Institute, and their colleagues. Within the University, working with a large number of research fellows and students has enabled my ideas in the subject to be straightened out, and their experimental results have been used in the book. These include Ian Barker, Annamaria Basile, David Breen, Mark Burford, Mireille Caralp, Judith Carroll, Nick Cotton, Susan Coleby, Catherine Cowey, Andrea Decker, Noreen Din, William Gaskill, Isabelle

Gélébart, Darren Heaton, Saad Jafar, Thomas Johnson, Jacob Kithinji, Naila Malak, Scott Oakes, Safar Öczan, Kate Pople, Mark Robson, Mark Rayner, Kemal Sangun, Gavin Shilstone, Harold Vandenberg, David Walker, and many others.

I would also like to thank my colleagues at Express Separations Ltd, a University spin-off company set up to transfer supercritical fluid technology to industry, and also the companies who have given Express research contracts. The companies would prefer to remain anonymous, and the work cannot be reported, even though some of it would provide useful examples for this book, and the work has been responsible for developing our understanding. Recently, Express has become associated with Separex Chimie Fine of Nancy and we have benefited from the extensive knowledge of its proprietor, Michel Perrut.

Finally, I would like to thank my wife Anna and the rest of the family for their indulgence during the production of this volume, and apologize for my pre-occupation with it.

I am writing this from Leeds, where 225 years ago Joseph Priestly worked on carbon dioxide, which he collected from a local brewery. He carried out experiments on photosynthesis and, equally important, obtained a medal for inventing soda water.

Leeds
February 1998   A.A.C.

Contents

Notation

a	radius of particle or chromatographic column
a, a_{ij}	van der Waals and Peng–Robinson parameters
a_i	liquid extraction parameter $= F_x\, S_i$
a_m	activity in mobile phase
a_{st}	activity in stationary phase
a, b, c, A, A', A''	parameters in empirical solubility equation
A	Helmholtz function, area, total area of matrix particles
A, B, C	constants and parameters
A, B, C, C_{st}, C_m	constants in van Deemter equation
$A[I, J, K]$	molar flow rate relative to that of fluid
b, b_i	van der Waals and Peng–Robinson parameters
b_i, c_i	parameters in density programme for extraction
B, B_{ij}	second virial coefficient
B^*	second virial coefficient reduced by the pair-potential parameters
B_i	parameters in density programme for chromatography
B_2	derivative of $\ln(f_2/p^{\ominus})$ with respect to $\ln x_2$ at constant p and T
c	number of components
c, c_i	concentration of component
c, d	parameters in the vapour pressure equation
c'	surface concentration
c_f	concentration in the fluid
c_m	concentration in mobile phase
$c_m^{\ominus}$	standard concentration in mobile phase
c_{st}	concentration in stationary phase
$c_{st}^{\ominus}$	standard concentration in stationary phase
c_0	initial concentration in matrix
$c_{2,i}$	velocity of molecule i of component 2
C_n, C_n'	coefficients in rate equation for extraction with matrix effects
C_p	heat capacity at constant pressure

C_V	heat capacity at constant volume
d_p	particle diameter
d_{st}	thickness of stationary phase layer
D, D_i	diffusion coefficient
D_{st}	diffusion coefficient in stationary phase
D_{12}	binary diffusion coefficient
f	number of degrees of freedom, total feed flow rate relative to that of the fluid
f_i	fugacity coefficients of component i
f_{ij}	intermolecular force
F	force, volume flow rate
F_x	molar flow rate
F_1	molar flow rate of liquid solvent relative to that of fluid
G	Gibbs function
$\Delta G^{\ddagger}$	Gibbs function change for the formation of the activated complex
$\Delta G_{chr}^{\ominus}$	standard Gibbs function change for transfer to the stationary phase
$\bar{G}^{\ddagger\,\ominus}_p$	standard Gibbs function change for the formation of the activated complex
g, g_{ij}	pair correlation function
g_p	moments of concentration distribution in chromatographic column
h	Planck''s constant, solvation parameter, theoretical plate height
h_{min}	minimum plate height
H	enthalpy
$\Delta H_{chr}^{\ominus}$	standard enthalpy change for transfer to the stationary phase
$\Delta H_v^{\ominus}$ $\ominus$	enthalpy of vaporization
I	intercept of extraction curve
J_E	flux of energy
$J_{x,E}$	flux of energy in the x-direction
J_i	molar flux of component i
$J_{x,i}$	molar flux of component i in the x-direction
$J_{x,i}^m$	flux of component i in the x-direction in the mass-fixed reference frame
$J_{x,i}^v$	flux of component i in the x-direction in the volume-fixed reference frame
k, k_i	capacity factor
k_B	Boltzmann constant
k_{ij}	binary interaction parameter

k_i, k_P, k_t	rate coefficients for initiation, propagation, and termination, respectively
$k^{\ddagger}$	rate coefficient for breakdown of activated complex
k_1	desorption rate coefficient
k_2	readsorption rate coefficient
K	partition coefficient
K_w	ionic product for water
K_x	equilibrium constant, partition coefficient (ratio of mole fractions)
$K^{\ddagger \ominus}_p$	standard equilibrium constant for activated-complex formation
$\bar{K}^{\ddagger \ominus}_p$	pseudo standard equilibrium constant for activated-complex formation
l	length of tube or column
L	entry plate, film thickness
m	mass in matrix particle, mass of monomer unit
m_i	mass of component i
m_p	moments of concentration distribution averaged over column cross-section
m_0	initial mass in matrix particle, mass injected in chromatography
m_1, m_2, m_3	successive extracted masses
M	mass of the matrix, number of components
M'	mass of fluid passed
M_i	molar mass of component i
$<M>_n$	number-averaged molar mass
n	total number of moles
n°	initial total number of moles
n_i	number of moles of component i
n°_i	initial number of moles of component i
N	number of theoretical plates
N, N_i	number of molecules
N_A	Avogadro"'s constant
p	pressure, number of phases
$p^{\ominus} \ominus$	standard pressure, usually 101 325 Pa (1 atmosphere)
p_c	critical pressure
p_{chr}	pressure of chromatography
p_i	partial vapour pressure of component i
p°_i	vapour pressure of the pure component i
p_n, q_n	extraction rate coefficients for extraction with matrix effects
p_r	pressure reduced by critical pressure

p_{ref}	reference pressure, typically 1 bar
p_v	vapour pressure
P	polydispersity
$P(r)\,dr$	probability of molecule at a distance between r and $r + dr$
r	distance from centre, proportion carried from plate, reaction distance
r, r_{ij}	intermolecular distance
r^*	intermolecular distance reduced by the pair-potential parameters
r_0	equilibrium distance
R	chromatographic resolution
$R[I]$	reflux ratio for component i
s	number of product molecules formed from the activated complex
S	entropy, solubility in terms of mass per unit volume
$S_i, S[I], S[I,J]$	solubility in terms of mole fraction
$\Delta S_{chr}^{\ominus}$	standard entropy change for transfer to the stationary phase
t	time
t_c	characteristic time for extraction
$t_{c,2}$	correlation time for molecular velocities of component 2
t_M	mobile phase retention time
t_R	retention time
T	absolute temperature
T_c	critical temperature
T_r	temperature reduced by critical temperature
T^*	temperature reduced by the pair-potential parameters
u	dummy time variable
U	internal energy
v	small volume
v, v_x	velocity of mass-fixed with respect to earth-fixed reference frame
v_{mv}	velocity of the mass-fixed with respect to volume-fixed reference frame
v_{opt}	optimum mobile phase velocity
v_y	velocity in the y-direction
$v_0, v_{0,i}$	average velocity of fluid in chromatographic column
V	molar volume, volume of matrix
V'	volume of fluid passed

V, V_{ij}	intermolecular pair potential
V_c	critical molar volume
V_i	partial molar volume of component i
$V_i^{\ominus}$	partial molar volumes under ideal-gas conditions
V_m	molar volume of solute
V_M	mobile phase retention volume
V_r	volume reduced by critical volume
V_R	retention volume
ΔV	volume change on reaction
$\Delta V^{\ddagger}$	volume of activation
$\Delta \bar{V}^{\ddagger\ominus}$	volume of activation in an ideal gas
w_i	mass fraction of component i
$w_{\frac{1}{2}}$	peak width at half height
x_i	mole fraction of component i in the liquid or fluid phase
$X[2, J]$	mole fractions in the liquid phase
y_i	mole fraction of component i in the fluid or gas phase
$Y[I, J]$	mole fractions in the supercritical fluid phase
z	integral of time used in liquid extraction, longitudinal coordinate in column
z_i	mole fraction of oligomer i being extracted (not including the fluid)
Z	compression factor
Z_c	critical compression factor
α	distance parameter for concentration profile, chromatographic selectivity
β	stationary to mobile phase ratio
β, γ	parameters in van der Waals equation for competing reactions
γ	unknown factor in the van Deemter B term
δ	as prefix = elemental amount
Δ	as prefix = change of thermodynamic quantity for phase transfer or reaction
ε	initiator efficiency
$\varepsilon, \varepsilon_{ij}$	potential well depth
η	dynamic viscosity coefficient
κ_T	isothermal compressibility
λ	thermal conductivity
λ_n	roots of $\lambda\cot(\lambda) = 1 - ha$ or $\lambda\tan(\lambda) = ha$
$\mu_m^{\ominus}$	chemical potential in mobile phase at standard pressure and infinite dilution

ρ	density
ρ_c	critical density
ρ_{ref}	reference density, typically $700\,\mathrm{kg\,m^{-3}}$
σ, σ_{ij}	intermolecular pair potential distance parameter
τ	average time spent in mobile phase
ϕ	statistical or steric parameter
ϕ_i	fugacity coefficient of component i
ω, ω_i	acentric factor

1 *A single substance as a supercritical fluid*

1.1 Introduction

1.1.1 Phase diagram of a single substance

When two molecules approach each other in a fluid, at a temperature where their relative speed is likely to be low, their mutually attractive forces will bring about a temporary association between them. If there is a sufficient density of molecules, there is the possibility of condensation to a liquid. On the other hand, if the temperature and the probable relative speeds are high, the attractive force will be too weak to have more than a slight effect on the molecular velocities, and condensation cannot occur however high the molecular density. It is therefore reasonable to expect, on the basis of molecular behaviour, that for every substance there is a temperature below which condensation to a liquid (and evaporation to a gas) is possible, but above which these processes cannot occur. That there is a *critical temperature* above which a single substance can only exist as a fluid and not as either a liquid or gas was shown experimentally 170 years ago by Baron Charles Cagniard de la Tour (1822). He heated substances, present as both liquid and vapour, in a sealed cannon which he rocked back and forth and discovered that, at a certain temperature, the splashing ceased. Later he constructed a glass apparatus in which the phenomenon could be more directly observed.

These phenomena can be put into context by reference to Fig. 1.1, which is a phase diagram of a single substance. The diagram is schematic, the pressure axis non-linear and the solid phase at high temperatures occurs at very high pressures. (Further solid phases and also liquid crystal phases can also occur on a phase diagram.) The areas where the substance exists as a single solid, liquid, or gas phase are labelled; as is the triple point where the three phases coexist. The curves represent coexistence between two of the phases. If we move upwards along the gas–liquid coexistence curve, which is a plot of vapour pressure versus temperature, both temperature and pressure increase. The liquid becomes less dense because of thermal expansion and the gas becomes more dense as the pressure rises. Eventually, the densities of the two phases

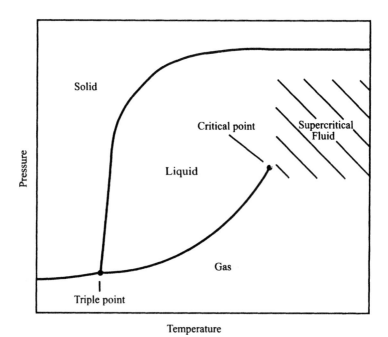

Fig. 1.1   The phase diagram of a single substance.

Table 1.1   Substances useful as supercritical fluids, with parameters from Reid et al. (1987)

Substance	Critical temperature, T_c (K)	Critical pressure, p_c (bar)	Critical compression factor, Z_c	Acentric factor, ω
Carbon dioxide	304	74	0.274	0.239
Water	647	221	0.235	0.344
Ethane	305	49	0.285	0.099
Ethene	282	50	0.280	0.089
Propane	370	43	0.281	0.153
Xenon	290	58	0.287	0
Ammonia	406	114	0.244	0.250
Nitrous oxide	310	72	0.274	0.165
Fluoroform	299	49	0.259	0.260

become identical, the distinction between the gas and the liquid disappears and the curve comes to an end at the *critical point*. The substance is now described as a fluid. The critical point has pressure and temperature coordinates on the phase diagram, which are referred to as the critical temperature, T_c, and the critical pressure, p_c, and which have particular values for particular substances, as shown by example in Table 1.1.

The disappearance of the distinction between the liquid and gas phases can be graphically illustrated by conducting a modern version of the Cagniard de la Tour experiment in which the meniscus between a liquid and a gas in a view cell disappears at the critical temperature. Figure 1.2 shows three schematic representations of a view cell in which this experiment is conducted at appropriate points on the liquid–gas coexistence curve. Cell (a) is at the lowest temperature and shows the liquid and gas phases with a meniscus between them. As the temperature and pressure rise and the density difference between the two phases becomes less, the meniscus becomes less distinct, as shown in cell (b). In practice the meniscus is no longer flat, because of temperature fluctuations and the small density difference. When the critical point is passed the meniscus disappears altogether, as shown in cell (c).

In recent years, fluids have been exploited above their critical temperatures and pressures and the term supercritical fluids has been coined to describe these media. The greatest advantages of supercritical fluids occur typically not too far above (say within 100 K of) their critical temperatures. Nitrogen gas in a cylinder is a fluid; however, it is not usually considered as a supercritical fluid, but is more often described by an older term as a permanent gas. The region for supercritical fluids is the hatched area in Fig. 1.1. It has been shown to include a region a little below the critical pressure as processes at these conditions are sometimes included in discussions as 'supercritical'. Lower pressures are also important in practice because these conditions are relevant to separation stages in supercritical processes. There are no phase boundaries below and to the left

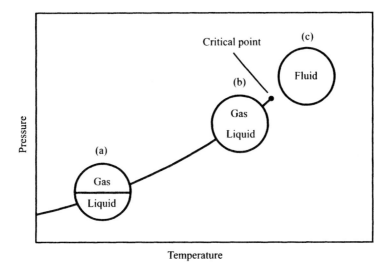

Fig.1.2   Disappearance of the meniscus at the critical point.

of the supercritical region in Fig. 1.1 and behaviour does not change dramatically on moving out of the hatched area in these directions. The liquid region to the left of the supercritical region has many of the characteristics of supercritical fluids and is exploited in a similar way. For this reason some people prefer the term near-critical fluids and the adjective subcritical is also used. The term supercritical fluid has, however, gained currency, is convenient and not a problem provided too rigid a definition is not applied. When two-component systems are considered in the next chapter, it will be seen that the definition is even more arbitrary. As will be seen below, supercritical fluids exhibit important characteristics such as compressibility, homogeneity, and a continuous change from gas-like to liquid-like properties. These properties are characteristic of conditions inside the hatched area in Fig. 1.1 and to different degrees in the area around it.

Table 1.1 shows the critical parameters of some of the important compounds useful as supercritical fluids. One compound, carbon dioxide, has so far been the most widely used, because of its convenient critical temperature, cheapness, chemical stability, non-flammability, stability in radioactive applications, and non-toxicity. Large amounts of CO_2 released accidentally could constitute a working hazard, given its tendency to blanket the ground, but hazard detectors are available. It is an environmentally friendly substitute for other organic solvents. The CO_2 that is used is obtained in large quantities as a by-product of fermentation, combustion, and ammonia synthesis and would be released into the atmosphere sooner rather than later, if it were not used as a supercritical fluid. Its polar character as a solvent is intermediate between a truly non-polar solvent such as hexane and weakly polar solvents. Because the molecule is non-polar it is often classified as a non-polar solvent, but it has some limited affinity with polar solutes because of its large molecular quadrupole. To improve its affinity with polar molecules further, CO_2 is sometimes modified with polar entrainers, as is discussed in the next chapter. However, pure CO_2 can be used for many organic solute molecules even if they have some polar character. It has a particular affinity for fluorinated compounds and is useful for working with fluorinated metal complexes and fluoropolymers.

CO_2 is not such a good solvent for hydrocarbon polymers and other hydrocarbons of high molar mass. Ethane, ethene, and propane become alternatives for these compounds, although they have the disadvantages of being hazardous because of flammability and of being somewhat less environmentally friendly. However, small residues of lower hydrocarbons in foodstuffs and pharmaceuticals are not generally considered a problem. Water has good environmental and other advantages, although its critical parameters are much less convenient and it gives rise to corrosion problems. Supercritical water is being used, at a research level, as a medium for the oxidative destruction of toxic waste. There is particular interest in both supercritical and near-critical water because of the behaviour of its polarity and this is discussed in a section at the end of this chapter. Ammonia has similar behaviour, is often considered and discussed, but not often used. Many halocarbons have the disadvantage of cost or of being environmentally unfriendly. Xenon is expensive, but is useful for small-scale experiments involving spectroscopy because of its transparency in the infrared, for example.

Although often pursued in practice for environmental reasons, the more fundamental interest in supercritical fluids arises because they can have properties intermediate between those of typical gases and liquids. Compared with liquids, densities and viscosities are less and diffusivities greater. The conditions may be optimum for a particular process or experiment. Furthermore, properties are controllable by both pressure and temperature and the extra degree of freedom, compared with a liquid, can mean that more than one property can be optimized. Any advantage has to be weighed against the cost and inconvenience of the higher pressures needed. Consequently, supercritical fluids are exploited in particular areas.

1.1.2 Pressure–volume–temperature behaviour of a single substance

Greater insight into the fluid behaviour of a single substance can be obtained by considering its pressure, p, as a function of molar volume, V, at constant temperature, T. These isotherms were first observed about 130 years ago for CO_2 by Thomas Andrews, working in the north of Ireland. The behaviour is shown schematically in Fig. 1.3 for any single substance. At the highest temperature, $T \gg T_c$, the isotherm appears on the figure only at high molar volumes and the curve approximates to the form $p \propto 1/V$, i.e. that of a perfect gas. As the temperature is lowered, an inflexion in the isotherms becomes perceptible, where $(\partial^2 p/\partial V^2)_T = 0$, as is shown in the isotherm for $T > T_c$. As the temperature is decreased further, the slope at the point of inflexion also decreases until it becomes zero at $T = T_c$. The point at which this happens is the critical point, i.e.

$$(\partial^2 p/\partial V^2)_T = (\partial p/\partial V)_T = 0 \quad \text{at the critical point.} \quad (1.1)$$

The isothermal compressibility, κ_T, defined as being equal to $-(1/V) \times (\partial V/\partial p)_T$, thus tends to infinity at the critical point. κ_T gives the rate of fractional change in volume with pressure at constant temperature. In the critical region κ_T is high, and supercritical and near-critical fluids are characterized as being highly compressible.

At lower temperatures still, for $T < T_c$, the isotherm is found experimentally to have three segments. The upper segment represents liquid, and the pressure falls rapidly as the volume increases. This connects with a segment, representing liquid and gas phases, which is at constant pressure; the vapour pressure, p_v.

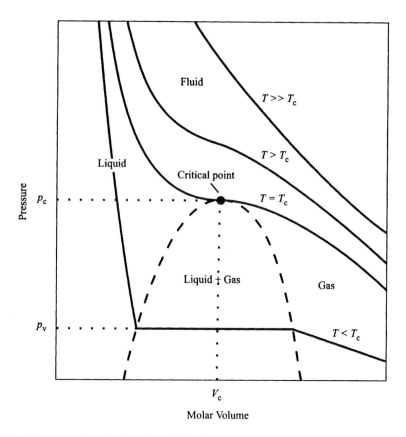

Fig.1.3 Pressure–volume isotherms for a single substance.

The final segment is for the gas phase and pressure falls more slowly with volume. The diagram is divided into areas. Within the dashed line there is a liquid + gas region; to the left of it a liquid region; and to the right of it a gas region. Above the critical isotherm there is a fluid or supercritical-fluid region.

The molar volume at the critical point, V_c, is a further critical parameter for a particular substance. For CO_2, for example, its value is $9.4 \times 10^{-5}\,m^3\,mol^{-1}$. This also determines a critical density, ρ_c, which for CO_2 is $10.6 \times 10^3\,mol\,m^{-3}$ or $466\,kg\,m^{-3}$. This critical density is about half the density of liquid CO_2 in a laboratory cylinder at $15\,°C$ and $55\,bar$. The compression factor, Z, (sometimes called the compressibility factor) is defined by $Z = pV/RT$, where R is the universal gas constant. Values of Z at the critical point, $Z_c = p_c V_c/RT_c$, are around 0.28 for many substances, as can be seen for the examples in Table 1.1, except for highly polar substances like water and ammonia, where they are lower.

1.2 Molecular effects

1.2.1 The intermolecular pair potential

This chapter began by discussing the existence of the critical temperature in terms of the interaction energy between molecules. The intermolecular pair potential, $V(r)$, is the potential energy of two molecules as a function of the distance r between them relative to their potential energy when $r \to \infty$. This potential will be dependent also, for non-spherical molecules, on the orientation of the molecules. However, this section is restricted to a discussion of spherically symmetrical molecules, which in some circumstances is approximately applicable to non-spherical molecules. A schematic diagram is shown in Fig. 1.4. As r decreases from large values, $V(r)$ become negative eventually reaching a minimum of $-\varepsilon$, the potential well depth, at r_0, the equilibrium distance. This first portion of the curve has a positive slope, and the force between the molecules, defined as being negative for attraction and equal to $-\mathrm{d}V(r)/\mathrm{d}r$, is attractive. As r decreases below r_0, the curve rises sharply with

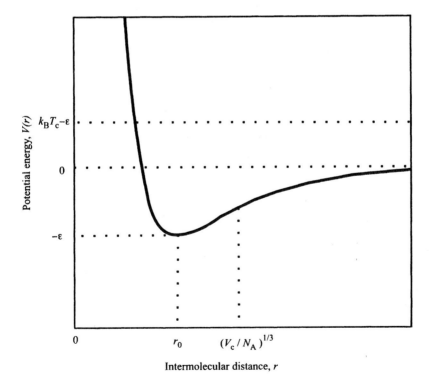

Fig.1.4 The intermolecular pair potential function.

a negative slope and represents a positive, repulsive force. The value of r as $V(r)$ passes through zero is a distance parameter of the pair potential, σ.

The study of intermolecular pair potentials has now reached a sophisticated level (Maitland *et al.* 1981). Earlier studies concentrated on fitting experimental data to simple analytical functions and these are still be used for illustrative purposes and for making simple estimates. An important example of a simple spherically symmetric functional form is the Lennard-Jones (12:6) potential which is given by

$$V(r) = 4\varepsilon \left[\left(\frac{\sigma}{r} \right)^{12} - \left(\frac{\sigma}{r} \right)^{6} \right] \tag{1.2}$$

The $(\sigma/r)^{12}$ term is the repulsive part of the potential and the $(\sigma/r)^{6}$ term the attractive part of the potential. It can be seen by inspection that $V(r) = 0$ when $\sigma = 0$ and by differentiation it can be shown that the minimum value of $V(r)$ is equal to $-\varepsilon$. It can also be readily shown that the distance corresponding to the potential minimum, r_0, is equal to $(2)^{1/6}\sigma = 1.122\sigma$.

Parameters for the intermolecular pair potentials for some of the less polar substances used as supercritical fluids, obtained from a variety of sources and meant to be only an indication of their values, are given in Table 1.2, with ε shown as a ratio with k_B, the Boltzmann constant. As expected, both ε and σ increase with the size of the molecule. Also shown are dimensionless ratios involving both the intermolecular potential parameters and the critical parameters. $k_B T_c/\varepsilon$ is seen to have a ratio of more than unity at about 1.4. $k_B T_c - \varepsilon$ is shown schematically on Fig. 1.4 to illustrate that the well depth is less than the average classical energy of vibration, $k_B T_c$, at the critical temperature of any physical bond that could be formed between the two molecules. This reinforces the assertion, made earlier, that above the critical temperature molecular energies are on average too high to allow the possibility of condensation.

The ratio $(V_c/N_A)^{1/3}/\sigma$ is also above unity, where N_A is Avogadro's constant. The quantity $(V_c/N_A)^{1/3}$ is the length of a cubic space occupied by a single molecule and is higher than σ and also higher than r_0, which is equal to 1.122σ

Table 1.2   Intermolecular parameters and related quantities

Substance	ε/k_B (K)	σ (pm)	$k_B T_c/\varepsilon$	$(V_c/N_A)^{1/3}/\sigma$
Carbon dioxide	246	375	1.21	1.43
Ethane	230	442	1.33	1.41
Ethene	205	423	1.38	1.42
Propane	237	512	1.56	1.35
Xenon	222	392	1.35	1.49
Nitrous oxide	232	383	1.34	1.43

for the Lennard–Jones potential and will have similar values for more realistic potentials. This situation is also shown on Fig. 1.4 and indicates that at the critical point intermolecular distances are above the distance corresponding to the minimum of the intermolecular potential well.

1.2.2 Structure in a fluid

We now consider the probability, $P(r)dr$, of a single molecule in a fluid having, on average, another molecule at a distance between r and $r + dr$, i.e. within a spherical shell of radius r and thickness dr, as illustrated in cross-section below.

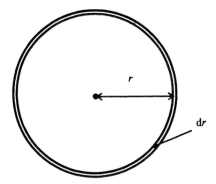

The average used here is over a large system approaching infinite size. For real molecules it is necessary to define the centre of a molecule, and it will be assumed here that this has been done in particular cases.

For a fluid in which the molecules are infinitely small and randomly positioned, such as a perfect gas, a value of $P(r)dr$ is readily available. The volume of the sphere is $4\pi r^2 dr$ and the density of molecules is V/N_A and therefore

$$P(r)dr = 4\pi(V/N_A)r^2\,dr \qquad (1.3)$$

For a real fluid, intermolecular forces will bring some structure and eqn (1.3) will not apply. This effect is taken into account by introducing an additional factor, the *pair correlation function*, $g(r)$, so that the equation becomes

$$P(r)dr = 4\pi(V/N_A)g(r)r^2\,dr \qquad (1.4)$$

The pair correlation function, sometimes called the radial distribution function, will have a value of zero for $r \rightarrow 0$, because the coincidence of molecules will not be allowed by repulsion. Repulsive forces, i.e. the size of the molecule, will also cause $g(r)$ to be extremely small for small values of r. At very high values of r, $g(r)$ will tend towards unity, i.e. $P(r)$ will tend to that of a perfect gas, because in a fluid there will be little correlation between the positions of two

well-separated molecules. At intermediate distances there is some structure
and this is illustrated by examples of pair correlation functions for different
densities in Fig. 1.5. The situation can be summarized by the statement that in a
fluid there is short-range order and long-range disorder.

The schematic curves in Fig. 1.5 are for a constant temperature, say 10 K
above the critical temperature. At high densities, $\rho > \rho_c$, the pair correlation
function is similar to that of a liquid. Outside the repulsive range it rises rapidly

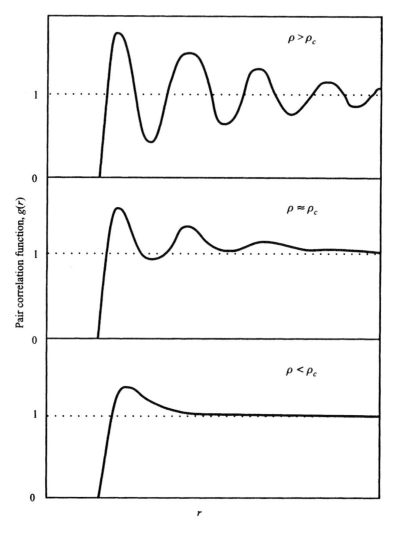

Fig.1.5   Pair distribution functions.

to a peak, which represents a first shell of surrounding molecules. The repulsive forces of these molecules then causes $g(r)$ to fall to a minimum before rising to another peak, corresponding to another shell of molecules. This repeats with each successive peak becoming less pronounced, because of the random nature of the successive shells, and the function converges to unity in an oscillatory way for high r. As the density is lowered the function becomes less distinct and around the critical density has the form shown in the middle curve in Fig. 1.5. Now the minima are shallow and there is a long tail to the function towards higher values of r, indicating a high correlation length. Such curves have not been observed experimentally, but have been obtained by calculation (Cochran and Lee 1989). At much lower densities, where only bimolecular interactions are significant, $g(r)$ has only one peak and as the density tends to zero

$$g(r) \rightarrow \exp[-V(r)/k_B T] \tag{1.5}$$

1.2.3 Critical opalescence and fluctuations

In the region around the critical point a fluid scatters visible light; this phenomenon being known as critical opalescence. The effect shows that there are significant density fluctuations over distances equal to visible-light wavelengths, which are of the order of 10^3 molecular diameters. In simplistic terms, the gas can be considered as consisting of large clusters of molecules with regions of lower density between. It can be thought of as an approach to the condensation behaviour possible below the critical temperature. This behaviour is more objectively described by the long tail in the pair correlation function around the critical density, as shown in Fig. 1.5.

A simple explanation of why there are higher density fluctuations in the critical region can be made by considering two vessels containing a fluid connected by a tube, as shown below, which are a little above the critical temperature.

The densities in both vessels are close to the critical density; in the right-hand vessel a little below the critical density and in the left-hand vessel a little below. However, because $(\partial p/\partial V)_T$ is very small under these conditions, the pressures in the two vessels will be very close, so there will be only a small tendency for the densities to equalize by fluid flowing through the connecting tube.

For a more rigorous approach a large amount (approaching an infinite amount) of a single substance at equilibrium is considered. It is then imagined that the substance is divided into a large number of small systems of equal volume, v, so that molecules can pass between the systems. As the substance is at equilibrium, the temperature and the chemical potential in each of the systems will be the same. This collection of systems is known as a *grand canonical ensemble*. Let a particular system contain N molecules and the number of molecules averaged over all the systems in the ensemble be given by $\langle N \rangle$. The quantity $N - \langle N \rangle$ is the difference between the number in a particular system and the mean number and is known as the fluctuation of N. Its value averaged over the ensemble will be zero, but the mean squared fluctuation $\langle (N - \langle N \rangle)^2 \rangle = \langle N^2 \rangle - \langle N \rangle^2$ will be finite. The methods of statistical mechanics, as given, for example, in Tolman (1979) or in a simpler introductory textbook by Turner and Betts (1974) can be used to show that

$$\frac{\langle N^2 \rangle - \langle N \rangle^2}{\langle N \rangle^2} = \frac{k_B T \kappa_T}{v} \tag{1.6}$$

Thus the mean squared fluctuations of the number of molecules, as a fraction of the square of the mean number of molecules in the system is inversely proportional to the volume, v. It is also proportional to the isothermal compressibility, κ_T, which is large in the region around the critical point.

There is also a relationship between the fluctuations and the pair distribution function (Kirkwood and Buff 1951), which is

$$\frac{\langle N^2 \rangle - \langle N \rangle^2}{\langle N \rangle^2} = \frac{4\pi}{v} \int_0^\infty r^2 \{g(r) - 1\}\, dr + \frac{1}{\langle N \rangle} \tag{1.7}$$

The second term on the right-hand side of eqn (1.7) is usually small and again there is a factor of $1/v$ in the integral term. The long tail in $g(r)$ for conditions near the critical point, shown in Fig. 1.5, will enhance the value of the integral in eqn (1.7) and corresponds to large fluctuations in density in the critical region.

1.3   Equations of state

We have so far discussed five thermodynamic functions of state: pressure, p; molar volume, V; temperature, T; isothermal compressibility, κ_T; and compression factor, Z. Later another seven will be used: internal energy, U; enthalpy, H; heat capacity at constant volume, C_V; heat capacity at constant pressure, C_p; entropy, S; Helmholtz function, A; and Gibbs function, G. The state of a system at equilibrium is defined by any two of these, most commonly V and T or p and T. Thus if V and T are known for a particular system, the remaining functions of state are defined. It is therefore possible in principle to write relationships for any function in terms of any two others, for example for the pressure in terms of volume and temperature:

$$p \equiv p(V, T) \tag{1.8}$$

This equation is the most common example of an *equation of state*. For a perfect gas, the equation can be written more specifically as $p = RT/V$. Once a relationship has been accepted as reflecting the behaviour of the system for three of the functions of state, the relationships between any three of the other functions of state are defined, apart from arbitrary (integration) constants in some cases. Thus acceptance of a particular equation of state allows all the thermodynamic functions to be calculated, using standard thermodynamic relationships.

Although equations of state are often presented with pressure as the calculated variable as in eqn (1.8), particularly for simple analytical forms, it is also common to give them as equations for Z and A. Equations for Z often have simple analytical forms; for example, $Z = 1$ is the equation of state for a perfect gas. Equations for A are used sometimes for complex numerical equations because they are convenient and accurate for the numerical calculation of the other functions and because the arbitrary zero of energy is included in the equations.

Although the equations of state that will be discussed in this chapter are fairly simple analytical forms, the most accurate equations are complex numerical forms that have been obtained by intelligent fitting of a wide range of thermodynamic data, such as is carried out at the International Union of Pure and Applied Chemistry Thermodynamic Tables Project Centre at Imperial College in London. They have carried out a study for a number of gases suitable as supercritical fluids and, in particular, for carbon dioxide (Angus *et al.* 1976). Their compilation, which contains the basic equation and a number of tables of functions under various conditions, is unfortunately out of print. A more recent equation of state for carbon dioxide is that published by Span and Wagner (1996). A large amount of work has been carried out for water, because of its importance in power generation and in geological studies, much of it under the auspices of the International Association for the Properties of Water and Steam (IAPWS). They authorize reports of work on data correlation which appear typically in the *Journal of Physical and Chemical Reference Data* and the work appears in compilations known as 'Steam Tables' (e.g. Haar *et al.* 1984). Many of the more complex equations of state are incorporated into physical property computer data packages, which are commercially available. For calculations connected with the publication of basic physical property data, it is important to use the best and latest thermodynamic data available. For many other purposes, the differences in values predicted by different equations of state are small enough not to be important, except near the critical point.

1.3.1 The virial equation of state

An equation for the pressure of a substance can be written, in terms of temperature and volume, in the following form of the *virial equation of state*

$$\frac{P}{RT} = \frac{A(T)}{V} + \frac{B(T)}{V^2} + \frac{C(T)}{V^3} + \frac{D(T)}{V^4} \cdots \tag{1.9}$$

The infinite series on the right-hand side of eqn (1.9) is found to converge above the critical temperature and is perfectly general, provided that the series converges, because the *virial coefficients* of the inverse powers of volume, $A(T)$, $B(T)$, etc., are arbitrary functions of temperature. However, because it is known from experiment that at very large volumes a substance tends to behave as a perfect gas, it must be true that $A(T) = 1$. It can be shown by the methods of statistical mechanics (the so-called Mayer's cluster expansion) that the second, third, etc. terms on the right hand side of eqn (1.9) arise, respectively, from bimolecular, termolecular, etc. interactions. In a situation in which bimolecular interaction cause appreciable deviations from perfect-gas behaviour, but the effect of termolecular and higher interactions are negligible, the substance is described as a *dilute gas*. In this situation

$$\frac{p}{RT} \approx \frac{1}{V} + \frac{B(T)}{V^2} \tag{1.10}$$

It can be easily shown for a dilute gas that the *second virial coefficient*, $B(T)$, is equal to the volume occupied by one mole of the gas less the volume that the gas would occupy at the same pressure and temperature if it were behaving as a perfect gas. $B(T)$ is negative at low temperatures, where attractive forces are important and positive at higher temperatures, where the effect of attractive forces is not significant and repulsive forces predominate. It is related to the spherically symmetric pair potential by

$$B(T) = -2\pi N_A \int_0^\infty r^2 \left\{ \exp[-V(r)/k_B T] - 1 \right\} dr \tag{1.11}$$

where N_A is Avogadro's number.

Equation (1.11) can be written in some cases in terms of variables B^*, r^* and T^*, which are reduced by the pair-potential parameters and defined by:

$$r^* = r/\sigma; \quad T^* = k_B T/\varepsilon; \quad B^* = 3B/2\pi N_A \sigma^3 \tag{1.12}$$

The Lennard–Jones pair potential function in the form of eqn (1.2) can be written in terms of r^* as

$$V(r) = 4\varepsilon \{ (r^*)^{-12} - (r^*)^{-6} \} \tag{1.13}$$

Substitution of eqns (1.12) and (1.13) into eqn (1.11) gives for the Lennard–Jones pair potential

$$B^*(T^*) = -3 \int_0^\infty (r^*)^2 \{\exp[-4\{(r^*)^{-12} - (r^*)^{-6}\}/T^*] - 1\} \, dr^* \quad (1.14)$$

The right-hand side of eqn (1.14) is only a function of T^* and can be integrated to give

$$B^*(T^*) = -\frac{\sqrt{2}}{4} \sum_{j=0}^{\infty} \left[\left(\frac{1}{T^*}\right)^{(2j+1)/4} \frac{2^j}{j!} \Gamma\left(\frac{2j-1}{4}\right) \right] \quad (1.15)$$

This function can be readily calculated and is available in tabular form in some textbooks (e.g. Maitland *et al.* 1981). If the Lennard–Jones parameters are known for a particular substance, the second virial coefficient can be calculated using eqn (1.15) or published tables, together with the definitions in eqns (1.12).

Although only really valid for a dilute gas, eqn (1.10) is sometimes used up to half the critical density. It is not of much use for pure substances as supercritical fluids, but is introduced here so that it can be extended later to a binary mixture for making estimates of solubilities at low pressures. These are relevant to separation stages in supercritical-fluid processes.

1.3.2   The law of corresponding states and the acentric factor

The reduced pressure, p_r, reduced volume, V_r, and reduced temperature, T_r, are defined by:

$$p_r = p/p_c; \quad V_r = V/V_c; \quad \text{and} \quad T_r = T/T_c \quad (1.16)$$

An equation of state can be written in the form

$$p_r \equiv p_r(V_r, T_r) \quad (1.17)$$

According to the law of corresponding states, the function $p_r(V_r, T_r)$ is a universal function for all substances. The law, which is of course only approximately true, is based on the fact that diagrams of p–V–T behaviour, such as Fig. 1.3, all have approximately the same shape, although the dimensions are different. Thus in principle the behaviour of any substance can be predicted from that of a suitable model substance, using the critical parameters of both substances. The principle also extends to all thermodynamic properties when these are reduced by suitable critical constants. The law predicts that Z_c is a constant, as can be seen to be approximately true from Table 1.1. However, the law of corresponding states does not make very accurate predictions, especially for non-spherical molecules, and has therefore been enhanced by introducing the concept of the *acentric factor*.

As it is supposed to hold in both the gas and the liquid phase, the law of corresponding states will predict the same vapour pressure curve for all substances in terms of reduced pressure versus reduced volume. This is found to be approximately the case for many substances whose molecules were spherically symmetric, and it is also found that their vapour pressure falls to approximately $0.1p_c$ when the temperature falls to $0.7T_c$. For the 'model fluid' illustrated by the dashed curve in Fig. 1.6, this is shown to be exactly the case, i.e. p_r falls from 1 to 0.1 when T_r falls from 1 to 0.7. For real fluids, especially those with non-spherically symmetric molecules, this is not the case and this is illustrated by the solid curve in the figure. Asymmetric molecules in a liquid rotate more freely as the temperature rises and for this to happen they must move further apart on average. When this happens their intermolecular binding energy is reduced and they pass more easily into the gas phase. Thus the vapour pressure will rise more rapidly with temperature for asymmetric molecules than for spherically symmetric molecules. Polar molecules will also lose attractive potential energy as the temperature rises and their orientation becomes more random, and this will cause a more rapid change in vapour pressure with temperature. This will be especially true when hydrogen bonding is involved.

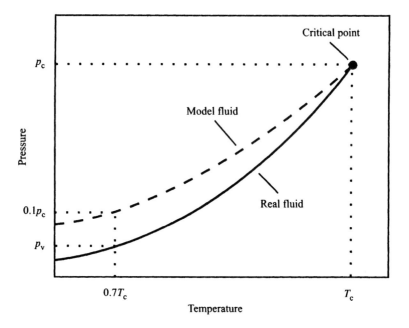

Fig.1.6 Vapour pressure curves to illustrate the acentric factor.

To quantify these effects an acentric factor, ω, has been defined by (Pitzer 1955)

$$\omega = -\log p_r(T_r = 0.7) - 1 \tag{1.18}$$

Thus for spherically symmetrical molecules, where $p_r(T_r = 0.7) \approx 0.1$, such as xenon, ω is essentially zero and for methane it is small at 0.011. Values for some other substances are shown in Table 1.1.

In the *extended law of corresponding states* equations such as (1.17) are modified to include the acentric factor as follows:

$$p_r \equiv p_r(V_r, T_r, \omega) \tag{1.19}$$

This allows more realistic predictions to be made using a model substance, such as methane, although difficulties are experienced with substances which form hydrogen bonds. This method forms the basis of some prediction methods used in computer packages.

1.3.3 The van der Waals equation of state

The van der Waals equation in its simplest form, as used in this book, is not very accurate in predicting the values of properties, but it is very useful for explaining principles and showing the qualitative behaviour of supercritical fluids. For a one-component fluid it is given by

$$p = \frac{RT}{V - b} - \frac{a}{V^2} \tag{1.20}$$

where a and b are constants, known as the van der Waals parameters. The equation is an adaptation of the perfect-gas equation of state in which the volume has been reduced by b, the so-called *excluded volume*, to allow for the physical size of the molecules, and the pressure increased by a/V^2, to allow for the effect of attraction between the molecules. Above the critical temperature and outside the liquid + gas region, the equation reproduces qualitatively the behaviour of the $p-V$ isotherms in Fig. 1.3. It is described as a cubic equation of state because, when multiplied throughout by $V^2(V-b)$ to remove both denominators, the equation contains a term in V^3.

Equation (1.20) is differentiated twice with respect to V at constant T to obtain expressions for $(\partial p/\partial V)_T$ and $(\partial^2 p/\partial V^2)_T$, and the critical values of V and T substituted into these expressions, which are then set equal to zero, in accordance with eqn (1.1). The two equations thus obtained, together with eqn (1.20) into which the critical values of p, V, and T have been substituted, are then solved to obtain relationships between the critical parameters and the van der Waals parameters, which are:

$$p_c = a/27b^2; \quad V_c = 3b; \quad \text{and} \quad T_c = 8a/27Rb \tag{1.21}$$

Note that V_c depends only on the excluded volume, and that T_c, which is known to depend on the depth of the pair potential well, is greater when the attractive parameter a is greater, but less when the molecules can approach each other less closely because of a greater value of b. The critical compression factor, $Z_c = p_c V_c / R T_c$ can be calculated from eqns (1.21) to be equal to 0.375, which is greater than the experimental values given in Table 1.1.

The inverse of eqns (1.21) can be used to calculate the van der Waals parameters a and b from the critical parameters for a particular substance but, as there are only two of these, one has to choose two of the three critical parameters to make the calculation. Because the van der Waals value of Z_c is unrealistic, the choice, for example, of p_c and T_c will lead to unrealistic values for V_c. For carbon dioxide, values of critical volume and density will be in error by some 50 per cent. As long as this is understood and an appropriate choice made as to which two critical parameters are used, qualitative modelling calculations can still be useful.

Substitution of critical parameters using eqns (1.21) for a and b in eqn (1.20) and then making further substitutions using the definitions of reduced quantities in eqn (1.16) gives the following form of the van der Waals equation:

$$p_r = \frac{8T_r}{3V_r - 1} - \frac{3}{V_r^2} \tag{1.22}$$

The right-hand side of eqn (1.22) is a universal function of reduced variables and thus the van der Waals equation conforms to the law of corresponding states.

1.3.4   The Peng–Robinson equation of state

A large number of more complex and realistic equations of state have been proposed and an example of these is now discussed, that of Peng and Robinson (1976), which is chosen because of its wide application in the field of supercritical fluids. The Peng–Robinson equation is one of a family of cubic equations of state developed from that of van der Waals, which contains the Redlich–Kwong and Soave equations as other important members. For the Peng–Robinson equation, the second term in the van der Waals equation is modified by making a a function of temperature and including b in the denominator of the second term, as shown below.

$$p = \frac{RT}{V - b} - \frac{a(T)}{V^2 + 2Vb - b^2} \tag{1.23}$$

If the same calculation at the critical point is carried out using eqn (1.1) as was previously described for the van der Waals equation, the following relationships are obtained, when a and b are calculated from the critical temperature and pressure:

$$a(T_c) = 0.45724R^2T_c^2/p_c; \quad \text{and} \quad b = 0.07780RT_c/p_c \qquad (1.24)$$

By the same method V_c is calculated to be $3.9514\,b$ and also $Z_c = 0.3074$, which is much closer to the experimental values than that obtained from the van der Waals equation, although still 11 per cent away from the experimental value for carbon dioxide.

The variation of a with T was obtained by Peng and Robinson by fitting to experimental hydrocarbon vapour pressures and obtaining the relationship

$$a(T) = a(T_c)\{1 + (0.37464 + 1.54226\omega - 0.26992\omega^2)(1 - T_r^{1/2})\}^2 \qquad (1.25)$$

and thus the acentric factor is introduced into the equation. On substitution of $a(T)$ and b by functions of the critical parameters in eqn (1.23) and after some manipulation the Peng–Robinson equation in terms of reduced variables is found to be

$$p_r = \frac{3.2529T_r}{V_r - 0.25308}$$
$$- \frac{4.83825\{1 + (0.37464 + 1.54226\omega - 0.26992\omega^2)(1 - T_r^{1/2})\}^2}{V_r^2 + 0.5062V_r - 0.06405} \qquad (1.26)$$

The right-hand side of this equation is a universal function of V_r, T_r, and ω, i.e. is of the form of eqn (1.19), and the Peng–Robinson thus conforms to the extended law of corresponding states.

1.4 Behaviour of thermodynamic functions

A supercritical fluid changes from gas-like to liquid-like as the pressure is increased and its thermodynamic properties change in the same way. Close to the critical temperature, this change occurs rapidly over a small pressure range. The most familiar property is the density and its behaviour is illustrated in Fig. 1.7. This shows three density–pressure isotherms and at the lowest temperature, 6 K above the critical temperature, the density change is seen to increase rapidly at around the critical pressure. As the temperature is raised, the change is less dramatic and moves to higher pressures. One consequence is that it is difficult to control the density near the critical temperature and, as many effects are correlated with the density, control of experiments and processes can be difficult. Other properties, such as enthalpy also show these dramatic changes near the critical temperature.

To illustrate the significance and behaviour of enthalpy and entropy for those not familiar with process engineering design, a simple supercritical-fluid

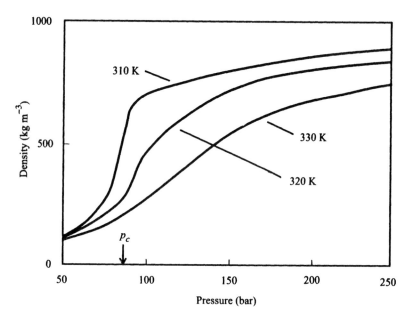

Fig.1.7   Density–pressure isotherms for carbon dioxide.

process is now discussed. The example chosen is an extraction with carbon dioxide in which the fluid is recycled and this is shown in Fig. 1.8, where the process is illustrated schematically as a graph of enthalpy versus entropy. Although extraction is taking place, for the purposes of illustration it will be assumed that the properties of the fluid are those of pure carbon dioxide. Point A represents liquid carbon dioxide in a condenser or storage vessel at $-3\,°C$ and 40 bar. It is pressurized to 200 bar to point B and then heated to $47\,°C$ at point C, where extraction occurs. The pressure is then reduced, with heating, to 40 bar at the same temperature for separation at point D. The carbon dioxide is then cooled again to $-3\,°C$ at point A for recycling.

During pumping, passing from point A to point B no heat is put into the fluid (ignoring friction), the entropy is therefore constant and the line A–B is vertical. The entropy of carbon dioxide at 270 K and 40 bar is $136\,J\,mol^{-1}\,K^{-1}$ and at 200 bar carbon dioxide has the same value of entropy at 280 K. Therefore the carbon dioxide has increased in temperature by 10 K due to the compression work done on it. At 270 K and 40 bar the enthalpy is $21.7\,kJ\,mol^{-1}$, whereas at 280 K and 200 bar it is $22.4\,kJ\,mol^{-1}$. The difference of $0.7\,kJ\,mol^{-1}$ is the work done by the pump and this value can be used, with the required flow rate to estimate the power output required for the pump. The carbon dioxide is still liquid and heat is required to bring it into the supercritical state between B and C before extraction. The enthalpy at 320 K

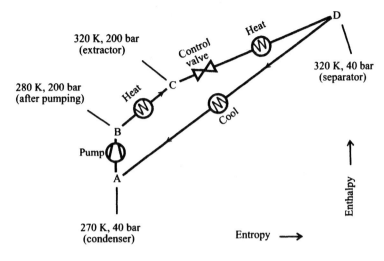

Fig. 1.8 Schematic diagram of an extraction process.

and 200 bar is $26.2\,\text{kJ}\,\text{mol}^{-1}$, and so $3.8\,\text{kJ}\,\text{mol}^{-1}$ is required from the heat exchanger. This value, with the flow rate, provides information for the design of the heat exchanger.

Following extraction, the pressure is reduced by a control value with heating. The enthalpy of carbon dioxide at 320 K and 40 bar is $34.8\,\text{kJ}\,\text{mol}^{-1}$ and so $8.6\,\text{kJ}\,\text{mol}^{-1}$ is required. This is the largest amount of heating required from a heat exchanger in the process and this is because on passing from C to D the fluid is changing from liquid-like to gas-like. Finally to condense the carbon dioxide to a liquid at 40 bar the enthalpy must be reduced to its value at A requiring $13.1\,\text{kJ}\,\text{mol}^{-1}$ to be removed by a cooler between D and A. With the flow rate, the size of the refrigeration unit needed can be calculated from this value. The amount of energy removed by the cooler equals, of course, the total of the compression and heat energy put into the system between A and D.

The heat capacity at constant pressure, C_p, shows dramatic behaviour because it is a derivative of the enthalpy. Figure 1.9 shows isotherms at 320 K for both heat capacities, C_V and C_p, of carbon dioxide. Although C_V shows only a small effect in the critical region, C_p goes through a peak which is three or four times higher than the surrounding values, even though the temperature is 16 K above the critical temperature. At constant pressure, the system expands when the temperature is raised. Work must be done against the intermolecular forces as the molecules move apart. Work must also be done against the atmosphere as the system expands. In the critical region the coefficient of thermal expansion at constant pressure is large, as can be seen from Fig. 1.7, and so the effect on C_p is also large.

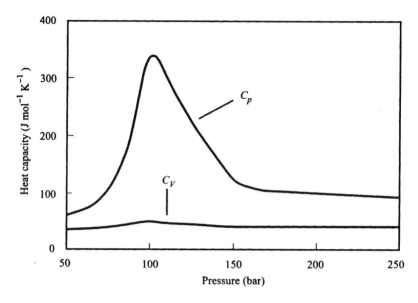

Fig.1.9   Heat capacities at 320 K for carbon dioxide.

1.4.1   Supercritical water

Water changes more dramatically when it becomes a supercritical fluid because of the breakdown in its structure as the temperature rises. The same is true for ammonia, and comments made in this section apply also to ammonia to a different degree. Liquid water has an extended structure due to intermolecular hydrogen bonding, but in the supercritical state its nature approaches that of a collection of light mobile molecules, which are iso-electronic with those of neon and have roughly the same molar mass. The high degree of association in liquid causes its dielectric constant (permittivity relative to vacuum) to be high at around 78 under ambient conditions, but as the temperature rises this falls dramatically, as shown in Table 1.3.

As a consequence, organic compounds have high solubilities in supercritical water, making extraction of and reactions with these compounds possible. In particular, there is considerable interest in the destruction of toxic organic compounds with molecular oxygen in supercritical water, as this process is found to go to the final oxidation products, such as carbon dioxide and water, relatively quickly. Conversely, salts are often much less soluble in supercritical than in liquid water and the ionic species present may also be different. Advantage can also be taken of the reduced dielectric constant below the critical temperature and there is much recent interest in sub-critical or 'superheated' water.

Table 1.3 Dielectric constant of water as a function of pressure, *p*, and temperature, *T* (Haar *et al.*1984)

T (°C)	*p* (bar)		
	300	400	500
0	89.2	89.6	90.1
100	56.4	56.8	57.1
200	35.9	36.3	36.6
300	22.0	22.6	23.1
400	6.0	10.5	12.2
500	1.7	2.3	3.4

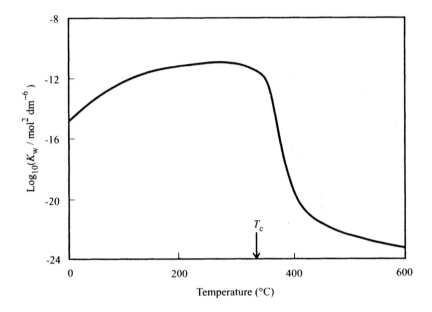

Fig.1.10   The ionic product, K_w, for water as a function of temperature at 250 bar.

A consequence of the lower dielectric constant of supercritical water is that it is a much less ionizing medium. This is illustrated in Fig. 1.10, which shows the logarithm of the ionic product for water, K_w, at 250 bar and as a function of temperature over a range including the critical temperature, T_c (Marshall and Franck 1981). As can be seen K_w first rises with temperature from its familiar value of 10^{-14} at 25 °C. However, as the temperature passes through T_c, K_w falls by many orders of magnitude as the density falls by a factor of 3 and the hydrogen-bonded structure breaks up.

2 Binary mixtures

2.1 Introduction

Binary mixtures will here be considered as a mixture of a fluid substance, such as those described in Chapter 1, which will be always called component 1, and a second substance, component 2. The binary mixtures that are important in supercritical fluids broadly fall into two areas, depending on the molar mass and critical temperature of component 2. In the first general area the second component has a molar mass and critical temperature mostly above, but not far above, those of component 1. In these cases the second component is added often in sizeable amounts to modify the solvation character of the fluid or, for example, to provide a reagent. Cases where the molar mass or the critical temperature of component 2 is lower than that of component 1 include the addition of a gas, such as hydrogen, to act as a reagent. The second area where consideration of binary mixtures is important is when involatile solutes of higher molar mass are dissolved in the fluid in the process of carrying out extraction, separation, or chemical reaction. The molar masses and critical temperatures of these second components will be typically much higher than that of component 1. In summary there are two areas of binary mixtures: in the first the component 2 is a modifier of relatively low molar mass; and in the second component 2 is a heavy solute.

The phase behaviour of binary mixtures like those described above also falls into two classes (Scott and van Konynenburg 1970; van Konynenburg and Scott 1980), and broadly speaking Class 1 phase diagrams describe fluid–modifier mixtures and Class 2 fluid–solute mixtures, although this is not always the case. Phase diagrams for binary mixtures are three-dimensional with p, T, and x axes, x_2 being the mole fraction of component 2. Two opposite faces of the three-dimensional diagram contain the phase diagrams of the individual components of the form of Fig. 1.1. In Class 1 there is a continuous *critical line* between the critical points of the two components, but in Class 2 the critical line is discontinuous. The concept of a critical line will be discussed below by example, but for present purposes they can be considered simply as features of the phase diagram.

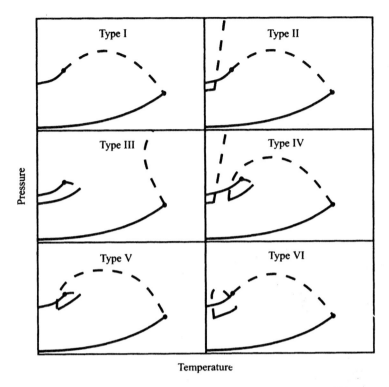

Fig. 2.1 Types of phase diagrams for binary mixtures projected on a $p-T$ plane: solid lines are either gas–liquid coexistence curves or projections of three-phase coexistence surfaces and dashed lines are critical lines.

These two classes are further divided into *types*, illustrated in Fig. 2.1 as projections on a $p-T$ plane (Rowlinson and Swinton 1982). In these diagrams, solid lines ending in bullets represent vapour pressure curves ending in critical points, other solid curves represent projections of three-phase boundaries and the dashed curves represent critical lines. The first five of these types of phase diagram were obtained using the van der Waals equation of state (Scott and van Konynenburg 1970; van Konynenburg and Scott 1980). Class 1 thus contains Types I, II, and VI (VI being added after the original Scott and van Konynenburg classification) and Class 2 Types III, IV, and V. As can be seen, the differences in general features between the types within one class involve liquid behaviour below the lower of the two critical temperatures and are therefore less relevant to supercritical-fluid behaviour. In this chapter, therefore, Type I behaviour will be discussed as an example of Class 1, typically relevant to modifiers, and Type III behaviour discussed as an example of Class 2, typically relevant to solutes. It should be noted that in a minority of texts the name class is used to describe what are here called types, but in this book original nomenclature of Scott and van Konynenburg is used.

Phase behaviour of binary mixtures over a wide range of temperature and pressure is a complex subject. The aim here is not to get involved too deeply in looking at the many types of behaviour possible, but to describe only in sufficient detail to explain the principles involved. Those wishing to take the subject further are referred to other textbooks including Rowlinson and Swinton (1982), Streett (1983), and McHugh and Krukonis (1994).

A further complication that should be referred to in this introduction is that, of course, real supercritical systems may contain more than one modifier and will probably have more than one heavy solute. Discussions of binary systems will give a qualitative understanding of the behaviour of multicomponent systems. The quantitative behaviour of binary systems is also of interest, although large changes in values may occur if another component is added even in small quantities. In fact it is a feature of compressible supercritical systems that they are sometimes very sensitive to changes in composition, as will be illustrated in the discussion of partial molar volumes below. An example is that the solubility of a compound is sensitive to, and typically enhanced by, the presence of other solutes in the fluid (Kurnik and Reid 1982). Treatment of multicomponent systems can be carried out using commercially available computer software designed for phase equilibrium studies.

2.1.1 Modifiers

Substances that are added to a fluid substance to change its solvent character are known as modifiers or entrainers, and the characteristics they impart include increased or decreased polarity, aromaticity, chirality, and the ability to further complex metal–organic compounds. Just as carbon dioxide is the most popular substance for use as a supercritical fluid, it is also the substance to which modifiers are most frequently added. This is because modifiers are seen as a way of making use of this desirable compound in circumstances where it is not the best solvent. For example, in the case of carbon dioxide, methanol is added to increase polarity, aliphatic hydrocarbons to decrease it, toluene to impart aromaticity, [R]-2-butanol to add chirality, and tributyl phosphate to enhance the solvation of metal complexes. They are often added in 5 or 10 per cent amounts by volume, but sometimes much more, say 50 per cent. They can have significant effects when added in small quantities and in these cases it may be the effect on surface processes rather than solvent character which is important. For example, the modifier may be effective in extraction by adsorbing on to surface sites, preventing the readsorption of a compound being extracted. Similarly, in chromatography, the modifier may cap active or unbonded sites on a stationary phase, preventing tailing of chromatographic peaks. A comprehensive review of modifiers has been made by Page *et al.*

Table 2.1 Substances which are useful modifiers in carbon dioxide, with critical parameters and acentric factors from Reid *et al.* (1987) (except for tributyl phosphate, where they are estimated from standard methods) and binary interaction parameters with carbon dioxide from Knapp *et al.* (1982) (except for methanol, where the value is an average from Table 2.2)

Substance	T_c (K)	P_c (bar)	ω	k_{12}
Methanol	513	81	0.556	0.110
Ethanol	514	61	0.644	
1-Propanol	537	51	0.623	
2-Propanol	508	48	0.665	
2-Butanol	536	42	0.557	
Acetone	508	47	0.304	
Acetonitrile	546	48	0.327	
Acetic acid	593	58	0.447	
Diethyl ether	467	36	0.281	0.047
Dichloromethane	510	63	0.199	
Chloroform	536	54	0.218	
Hexane	508	30	0.299	0.110
Benzene	562	49	0.212	0.077
Toluene	592	41	0.263	0.106
Tributyl phosphate	742	24	0.850	

(1992). Some compounds, which are commonly used as modifiers, are listed with their critical parameters and acentric factors in Table 2.1.

2.2 Phase behaviour

2.2.1 Type I behaviour

Although modifiers are deliberately different in character from the fluid substance to which they are added, and therefore the mixture will not be ideal, it is a convenient starting point to consider an ideal liquid mixture at a constant temperature below the critical temperature of both components, which obeys Raoult's law with its vapour behaving as a perfect gas. According to Raoult's law, the partial vapour pressure of component *i* above a mixture, p_i, is given by the product of its mole fraction in the mixture, x_i, and the vapour pressure of the pure component, p_i^o. The total pressure above the liquid mixture will be given by

$$p = x_1 p_1^o + x_2 p_2^o = p_2^o + (p_1^o - p_2^o)x_1 \tag{2.1}$$

This equation gives the relationship between pressure and mole fraction in the liquid and is linear. It is shown as the liquid line in Fig. 2.2. The mole fraction of component 1 in the vapour or gas phase, y_1, is given by the ratio of its partial pressure to the total pressure, i.e.

$$y_1 = \frac{x_1 p_1^o}{p_2^o + (p_1^o - p_2^o)x_1} \tag{2.2}$$

Substitution of y_1 for x_1 in eqn (2.1) using eqn (2.2) gives the relationship between pressure and mole fraction in the gas phase,

$$p = \frac{p_1^o p_2^o}{p_1^o + (p_2^o - p_1^o) y_1} \tag{2.3}$$

This relationship is plotted as the gas line in Fig. 2.2. Both lines form a phase diagram at constant temperature for the binary mixture, with a liquid region above the liquid line and a gas region below the gas line. Under any conditions represented by a point between these two lines, the system will form two phases with compositions given by the intersection of a horizontal line with the liquid and gas lines.

As previously mentioned, the fluid substance and the modifier substance will have somewhat different characteristics and the mixture will not be ideal. Typically both substances will have higher cohesive energy when pure than when mixed and the liquid line will exhibit a positive deviation from the Raoult's law line, as shown in Fig. 2.3. It has been shown typically without a maximum, which would give rise to positive azeotropy. This can occur, however, with extremely weak attraction between unlike molecules, when the phase diagram is described as Type IA, but will not be discussed further.

If the temperature of the system is raised and the critical temperature of the fluid substance is passed, the two-phase loop leaves the right-hand axis of

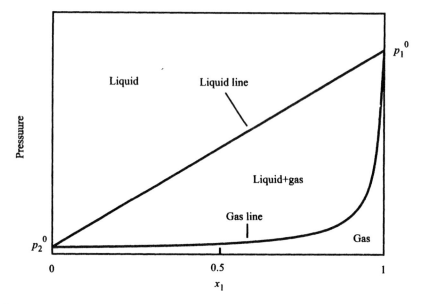

Fig. 2.2 An ideal liquid mixture at constant temperature.

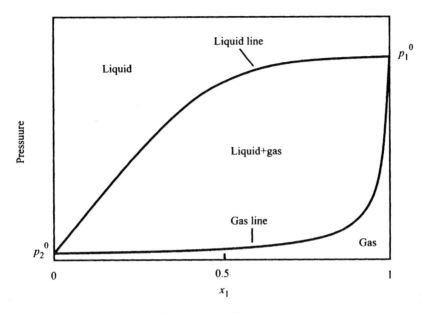

Fig. 2.3 A non-ideal liquid mixture of Type I at constant temperature.

the graph and the phase diagram takes on the form of Fig. 2.4, which a semi-schematic diagram for methanol–CO_2 at 50 °C (which uses the data of Brünner *et al.* 1987). As occurred below the critical temperature, at very low pressures a single gaseous phase exists at all compositions, which are mixtures of CO_2 and methanol vapour. At intermediate pressures, both gaseous and liquid phases can occur, dependent on composition. At high mole fractions of CO_2 the mixture is gaseous, at high methanol concentrations it is liquid, and at intermediate compositions both phases exist. The liquid + gas region reaches a maximum in pressure at the critical point (C) for this particular temperature. Consider what happens to a mixture of the critical composition at a pressure below the critical pressure (where it will be in two phases) as the pressure is raised. The liquid will dissolve more CO_2, the gas will solvate more methanol, and the gas will increase in density more rapidly than the liquid. Eventually, at the critical point, the compositions and densities of the two phases will become identical. Thus above the critical pressure only one fluid phase will exist. The critical point in this case is one point along a critical curve in the three-dimensional phase diagram, in which it will have coordinates of temperature, pressure, and composition, as will be seen below. For any mixture (for example, methanol–CO_2) at a given temperature (for example 50 °C) there will be a critical pressure (in the example 95.5 bar) and a critical composition (in the example mole fraction of $CO_2 = 0.84$). For a binary mixture, the term

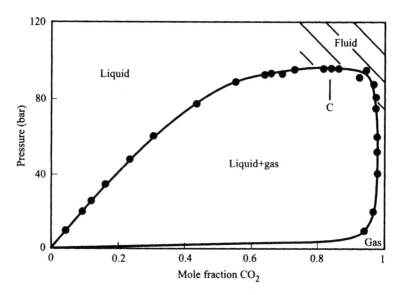

Fig. 2.4   Semi-schematic phase diagram of a methanol–CO₂ mixture at 50 °C, with experimental points from Brünner *et al.* (1987).

'supercritical' is more arbitrary than for a pure substance. Pure CO_2 is considered to be supercritical above 74 bar and the fluid rich in CO_2 above the two-phase envelope can also be considered to be a supercritical fluid, rather than a liquid. Hence the hatched area in the figure is that usually loosely called supercritical.

As the temperature rises further, the liquid + gas region becomes smaller and moves away from the right-hand CO_2 axis, and eventually disappears as the critical temperature of the fluid substance is reached. The overall situation is illustrated in Fig. 2.5, where the constant-temperature diagrams considered so far are seen as cross-sections of a three-dimensional temperature–pressure–composition phase diagram. The vapour pressure curve for the fluid substance is seen on the front face of the diagram, ending in its critical point, C_1, and that for the modifier substance is seen on the back face, ending in C_2. Joining the critical points is a 'critical line', which touches the maximum of the liquid + gas region at each temperature. Typical behaviour of this critical line is that for the system methanol–CO_2, where the critical line begins at C_1 at a pressure of 74 bar, and rises with temperature to 95.5 bar at 50 °C, then to 161 bar at 150 °C, after which it falls again to the critical pressure of methanol, 81 bar, at 239 °C. At the same time, the composition coordinate of the line moves progressively from pure CO_2 to pure methanol. The three-dimensional diagram of Fig. 2.5 is concisely represented as a projection of the two vapour pressure

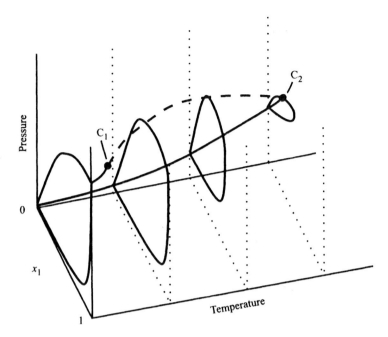

Fig. 2.5 A three-dimensional phase diagram of Type I, showing the vapour pressure curves for the pure components ending in the critical points C_1 and C_2, the critical line (dashed), and cross-sections at a number of temperatures.

curves and the critical line on to the front face of the diagram, as was previously shown in Fig. 2.1 (Type I).

Finally Fig. 2.6 shows a cross-section of a phase diagram of the form of Fig. 2.5 at constant composition. The region inside the *phase envelope*, and outside at lower temperatures, is a liquid region. At higher temperatures it is a gas and fluid region. The critical point shown is where the cross-section cuts the critical line and is not typically at the maximum of the phase envelope. As with Fig. 2.4, the distinction between what is described as 'liquid' and what is 'fluid' is arbitrary. In both diagrams, the more one moves towards the area described as liquid, the less compressible is the system, as will be described for a reaction medium in Section 8.2.1. Phase envelopes can be drawn and look similar for multicomponent systems.

2.2.2 Type III behaviour

This type of phase behaviour occurs for two components of more widely differing molar masses and critical temperatures, such as when an involatile liquid is mixed with a typical supercritical-fluid substance, such as a mixture of carbon dioxide and 1-hexadecanol. In this section solutes which have melting

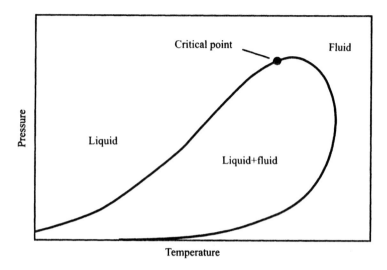

Fig. 2.6   The phase envelope of a binary mixture of Type I at constant composition.

points below the critical temperature of component 1 are considered. The existence of a solid solute phase is considered in the next section. Type III behaviour occurs when immiscibility which exists when both components are liquid extends above the critical temperature of component 1 and intersects with the critical line, as shown in projection earlier in Fig. 2.1 and in more detail in Fig. 2.7.

To help in understanding this figure, a temperature below the critical temperature of component 1 is considered, at say T_1 where both components are liquid. When liquids become immiscible, the liquid line in Fig. 2.3 is broken and a liquid–liquid region appears between its two arms. This is shown in Fig. 2.8, which is a cross-section of the three-dimensional phase diagram at T_1. As the pressure is lowered in the liquid–liquid region a gas or vapour will be produced at some point and this is shown by the horizontal liquid + liquid + gas line in Fig. 2.8. Along this line two liquid phases, each rich in one of the components, will be in equilibrium with a gas rich in component 1. The projection of this line on a p–T plane will be a point on the three-phase projection curve shown in Fig. 2.7.

When the system is taken just above the critical temperature of component 1, at say T_2 in Fig. 2.7, the cross-section changes to that in Fig. 2.9. The liquid + gas lobe rich in component 1 has now moved away from the right-hand axis and there is no longer a vapour pressure for component 1. The names of the phases inside and around this lobe are now rather arbitrary, as was the case in the Type I diagram. There is also a critical point, C, at the apex of this lobe

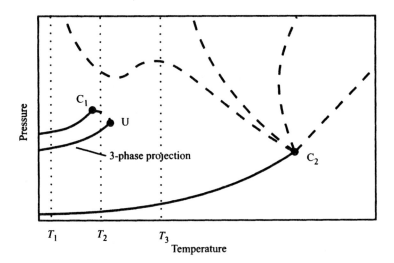

Fig. 2.7 A projection of a Type III phase diagram on a $p-T$ plane: alternative behaviour of the critical line emanating from C_2 is shown.

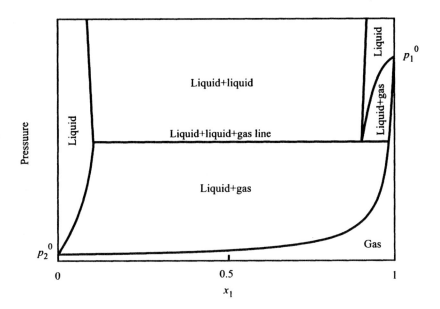

Fig. 2.8 A cross-section through the phase diagram represented by Fig. 2.7 at T_1.

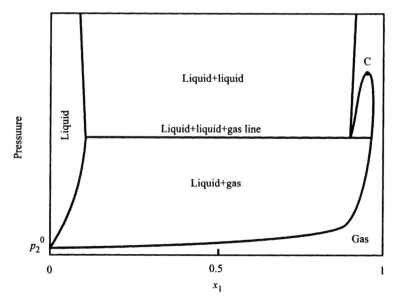

Fig. 2.9 A cross-section through the phase diagram represented by Fig. 2.7 at T_2, when there is not a critical line at high pressures.

similar to that in the Type I diagram, which is a point on the critical line which starts at the critical point of component 1. As the temperature is raised further, the lobe shrinks further and disappears, and at the same temperature the liquid + liquid + gas line also no longer exists. The projection of the liquid + liquid + gas lines on Fig. 2.7 therefore comes to an end at U and meets up with the critical line starting at C_1. Above the temperature of U, a cross-section through the three-dimensional phase diagram, at say T_3, may have the form of Fig. 2.10.

There is also a critical line which connects to the critical point of component 2, C_2, which can take various forms, depending on the nature of the components. Figure 2.7 shows some of the various types of behaviour possible by the critical line. The important criterion for work in supercritical fluids is whether, at the temperature of interest, there is a critical line at moderate pressures or not. If there is no critical line, or a critical line only at extremely high pressures, the cross-section of the phase diagram at the temperature of interest, say T_3, will take the form of Fig. 2.10. If a critical line exists at pressures not much above the pressures being used, the cross-section will take the form of Fig. 2.11. Complete miscibility then occurs at a high enough pressure. (It should also be mentioned that separation, so-called gas–gas separation, can occur at higher pressures in certain circumstances.) The right-hand portions of the curves in

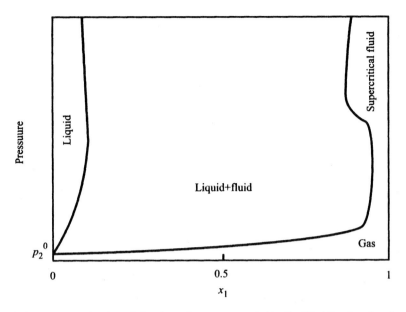

Fig. 2.10   A cross-section through the phase diagram represented by Fig. 2.7 at T_3, when there is not a critical line at high pressures.

Figs 2.10 and 2.11 are of importance as they are solubility curves. The behaviour of solubility will be discussed in the next chapter and these figures referred to.

2.2.3   Phase behaviour when the solute can be a solid

In many instances of importance component 2 is a solid above the critical temperature of component 1. In this case, the critical line is also discontinuous and can therefore be considered to be of Class 2, and the phase diagram has similarities to those of Type III systems, as shown in projection in Fig. 2.12. Around the critical point of component 1, the situation is the same as a Type III diagram, but as the solute becomes heavier the two curves of vapour pressure of component I and the projection of the liquid + liquid + gas surface lie very close together, as do the points C_1 and U. OT and TC_2 are the vapour pressure curves for the solid and liquid solute, respectively, ending in the solute critical point C_2. For temperatures where the vapour pressures of the pure component 2 are very small these are shown coincident with the zero-pressure axis. The point T is the triple point for the pure solute, and the line with positive slope emanating almost vertically from it is the solid + liquid line for the pure solute. In the presence of excess CO_2, however, melting to a solute-rich liquid phase will occur at pressures and temperatures to the right of the curve TL, which is

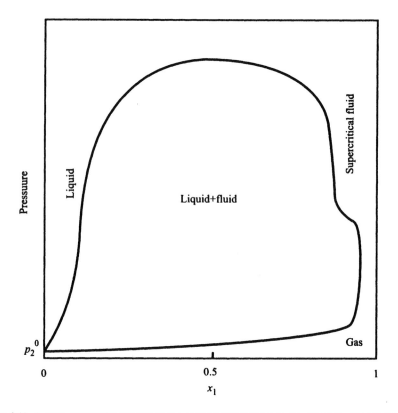

Fig. 2.11 A cross-section through the phase diagram represented by Fig. 2.7 at T_3, when there is a critical line at high pressures.

the projection of a three-phase region in which solid, liquid, and supercritical phases can coexist. TL is shown with a negative slope, but in some systems it has a positive slope and leans back towards the solid + liquid line. The dashed curve LC_2 is a critical line above which the liquid phase and the supercritical phase become identical and this ends in the critical end-point L.

Cross-sections of the three-dimensional $p-T-x$ diagrams, shown as $p-T$ projections in Fig. 2.12, are now considered. The part of the diagram of most interest to supercritical fluids are temperatures between any part of the critical line, UC_1, and the three-phase projection, TL, typified by T_1. This part of the diagram appears to be clear of features, but one should bear in mind that there are surfaces in it which do not appear in the figure, but are shown in cross-section. Figure 2.13(a) shows the cross-section at T_1. The features are qualitatively the same as those of Fig. 2.10 for a Type I diagram, but now many parts of the curves lie very close to the axes. For example, there is a line giving the composition of the solid phase in equilibrium with the fluid, which is

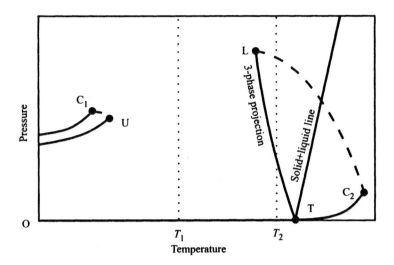

Fig. 2.12   Phase diagram of a binary mixture when component 2 is solid above the critical point of component 1, projected on a $p-T$ plane.

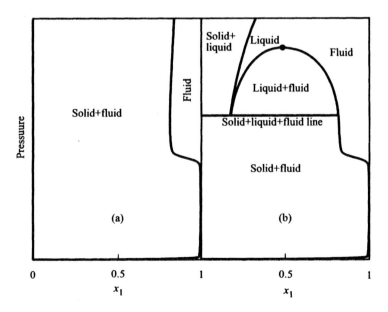

Fig. 2.13   Cross-sections through the phase diagram represented by Fig. 2.12, (a) at T_1 and (b) at T_2.

coincident with the $x = 0$ axis. In principle solids absorb the solvent, but this effect is normally small in extent. Similarly there is in principle a curve at very low pressures, representing the vapour of the solute in the presence of small amounts of the solvent, which is shown coincident with the zero-pressure axis. The only non-trivial feature of Fig. 2.13(a) is the curve at the right-hand side, which represents the supercritical solubility of the solute in the region of interest.

Figure 2.13(b) shows the cross-section at T_2 on Fig. 2.12, which intersects with the region of the phase diagram where the solutes form a liquid phase in the presence of component 1 under pressure. This situation is of importance where the solute has a melting point below or not too much above the temperature of interest. An extreme example is that of naphthalene, which melts at 357 K, but which can form a liquid phase under pressure in carbon dioxide as low as 340 K (McHugh and Paulitis 1980). Figure 2.13(b) is similar to Fig. 2.13(a) at low pressures, but at the pressure where the liquid region is intersected there is a horizontal line of solid + liquid + fluid coexistence, which projects as a point on the line TL in Fig. 2.12. At pressures above that line solid + liquid, liquid, liquid + fluid, or fluid exists, depending upon composition. At a higher pressure, the liquid and fluid phases become identical at a critical point, which is a point on the critical line LC_2 in Fig. 2.12.

2.3 Equations of state

2.3.1 The van der Waals equation for a binary mixture

The classes and types of phase diagram discussed above were developed using the van der Waals equation of state and it is useful to begin discussion with a quotation from van Konynenburg and Scott (1980): 'Although the van der Waals equation gives only a qualitative description of the thermodynamic properties of liquid mixtures, it rarely yields physical absurd results.' Introduced in Section 1.3.3, is most simply extended to mixtures by using the following combination rules for the parameters a and b,

$$a = a_{11}x_1^2 + 2a_{12}x_1x_2 + a_{22}x_2^2 \tag{2.4}$$
$$b = b_1x_1 + b_2x_2 \tag{2.5}$$

where a_{11}, a_{22}, b_1, and b_2 are the van der Waals parameters for the two pure components. a_{12} is a cross term for the two components and can be calculated in different ways, such as from a geometric mean, i.e.

$$a_{12}^2 = a_{11}a_{22} \tag{2.6}$$

Thermodynamic quantities for mixtures can be calculated from these equations using standard relationships. For example, an equation for the partial molar volume, V_i, the derivative of the total volume with respect to the number of moles of component i, can be found using the standard relationship

$$x_2 V_2 = (\partial V/\partial n_2)_{T,p} = V - (1 - x_2)(\partial V/\partial p)_{T,x}(\partial p/\partial x_1)_{T,V} \qquad (2.7)$$

The partial derivatives in eqn (2.7) are found, by substituting eqns (2.4) to (2.6) into eqn (1.20) and differentiating to give for component 2:

$$(\partial V/\partial p)_{T,x} = -[RT/(V - b)^2 - 2a/V^3]^{-1} \qquad (2.8)$$

$$(\partial p/\partial x_2)_{T,V} = RT(b_2 - b_1)/(V - b)^2 - 2[(a_{12} - a_{11})$$
$$+ x_2(a_{11} - 2a_{12} + a_{22}]/V^2 \qquad (2.9)$$

with a corresponding equation for component 1. For component 2, in the limit of $x_2 \to 0$, they are given by:

$$(\partial V/\partial p)_{T,x} = -[RT/(V - b_1)^2 - 2a_{11}/V^3]^{-1} \qquad (2.10)$$

$$(\partial p/\partial x_2)_{T,V} = RT(b_2 - b_1)/(V - b_1)^2 - 2(a_{12} - a_{11})/V^2 \qquad (2.11)$$

Important quantities in equilibria are the fugacity coefficients, f_i, which are defined here to be equal to px_i in the limit of zero pressure. As the derivative of chemical potential with respect to pressure at constant temperature and composition is equal to the partial molar volume, $RT[\partial \ln(f_i/p^{\ominus})/\partial p]_{T,x} = V_i$ and

Since $(dG = V dp)$
$\Rightarrow d\mu = V_i\, dp$

$$\ln(f_2/p^{\ominus}) = \frac{1}{RT} \int V_2\, dp + \text{constant} \qquad (2.12)$$

Substituting V_2 from eqn (2.7) and replacing dp by dV using the derivative $(\partial p/\partial V)_{T,x}$ gives:

$$\ln(f_2/p^{\ominus}) = \frac{1}{RT} \int \left[V\left(\frac{\partial p}{\partial V}\right)_{T,x} - (1 - x_2)\left(\frac{\partial p}{\partial x_2}\right)_{T,V} \right] dV + \text{constant}$$
$$(2.13)$$

After substituting for the partial derivatives in eqn (2.13) using eqns (2.8) and (2.9) and integrating, the following equation is obtained for f_2:

$$\ln(f_2/p^{\ominus}) = -\ln(V - b) + \frac{b_2}{V - b} - \frac{2[a_{12} + x_2(a_{22} - a_{12})]}{RTV} + \text{constant}$$
$$(2.14)$$

with a corresponding equation for f_1. At low pressure $\ln(f_2/p^{\ominus})$ becomes $\ln(px_2/p^{\ominus})$, V is large and the terms where it is contained in the denominator tend to zero, and $-\ln(V - b)$ becomes $\ln(p/RT)$. The constant therefore is equal to $\ln(RTx_2/p^{\ominus})$ and

$$\ln(f_2/p^{\ominus}) = \ln\frac{RTx_2}{(V-b)p^{\ominus}} + \frac{b_2}{V-b} - \frac{2[a_{12} + x_2(a_{22} - a_{12})]}{RTV} \qquad (2.15)$$

In the limit of infinite dilution as $x_2 \to 0$,

$$\ln(f_2/p^{\ominus}) = \ln\frac{RTx_2}{(V-b_1)p^{\ominus}} + \frac{b_2}{V-b_1} - \frac{2a_{12}}{RTV} \qquad (2.16)$$

The fugacity coefficient, ϕ_i, defined by $\phi_i = f_i/px_i$, is therefore given by

$$\ln\phi_2 = \ln\frac{RT}{p(V-b)} + \frac{b_2}{V-b} - \frac{2[a_{12} + x_2(a_{22} - a_{12})]}{RTV} \qquad (2.17)$$

and in the limit of infinite dilution by

$$\ln\phi_2 = \ln\frac{RT}{p(V-b_1)} + \frac{b_2}{V-b_1} - \frac{2a_{12}}{RTV} \qquad (2.18)$$

2.3.2 The Peng–Robinson equation for a binary mixture

The van der Waals equation has been put forward as a simple method of calculating qualitative trends. More complex cubic equations can give more quantitative predictions, with some reservations discussed below, and again the Peng–Robinson equation of state is given as an example because of its widespread use. The basic equations have been previously given as eqns (1.23) to (1.25), and for binary mixtures the parameters a and b are given by eqns (2.4) and (2.5) as for the van der Waals equation. However, a more complex equation defines the cross term a_{12}:

$$a_{12} = (a_{11}a_{22})^{1/2}(1 - k_{12}) \qquad (2.19)$$

where k_{12} is the *binary interaction parameter*. This parameter cannot be definitively predicted and must in general be found from experimental results. Some examples of values are given in Table 2.1, obtained by fitting selected experimental data (Knapp *et al.* 1982). A binary interaction parameter can also be used for the b parameters, but this will not be considered here. Using the methods described earlier for the van der Waals equation, the fugacity coefficient can be calculated to be given by

$$\ln\phi_2 = \ln\left(\frac{RT}{p(V-b)}\right) + \frac{b_2}{b}\left(\frac{pV}{RT} - 1\right) - \frac{a}{2\sqrt{2}RTb}$$
$$\times \left(\frac{2[a_{12} + x_2(a_{22} - a_{12})]}{a} - \frac{b_2}{b}\right) \ln\left(\frac{V + (1+\sqrt{2})b}{V + (1-\sqrt{2})b}\right) \qquad (2.20)$$

and in the limit of infinite dilution, the fugacity coefficient becomes

$$\ln \phi_2 = \ln\left(\frac{RT}{p(V - b_1)}\right) + \frac{b_2}{b_1}\left(\frac{pV}{RT} - 1\right) - \frac{a_{11}}{2\sqrt{2}RTb_1}$$
$$\times \left(\frac{2a_{12}}{a_{11}} - \frac{b_2}{b_1}\right) \ln\left(\frac{V + (1 + \sqrt{2})b_1}{V + (1 - \sqrt{2})b_1}\right) \qquad (2.21)$$

Equation (2.21) will be used in the next chapter to calculate solubilities in a supercritical fluid. Phase equilibria in general can be calculated using the Peng–Robinson equation, but it is possible to use software already prepared for this purpose. This is commercially available, and there is also a code published in Appendix B in McHugh and Krukonis (1994). Figure 2.14 shows three constant-temperature cross-sections through the phase diagram for methanol–CO$_2$ calculated from the Peng–Robinson equation of state using the parameters in Tables 1.1 and 2.1. It can be seen that the shape of the two-phase loop is different from that obtained experimentally and shown in Fig. 2.4. In particular, the experimental curve is much flatter at the top of the loop than the theoretical curve. This is also the case with other common analytical equations of state. New methods of constructing these phase diagrams are

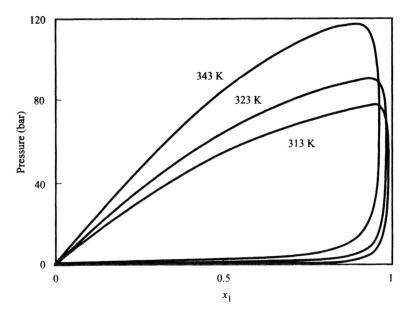

Fig. 2.14 Isothermal cross-sections through the phase diagram for methanol–CO$_2$ predicted by the Peng–Robinson equation.

Table 2.2 Binary interaction parameters, k_{12}, for methanol–CO_2 obtained by fitting the data of Brünner *et al.* (1987)	T (K)	Pressure (bar)	k_{12}
	323	50	0.068
	323	100	0.190
	373	50	0.110
	373	100	0.100
	373	150	0.074
	423	50	0.134
	423	100	0.125
	423	150	0.097

being developed to correct this discrepancy (Sengers 1994), but these are not yet in general use.

In principle, binary interaction parameters are constants, but in practice they can be made functions of temperature, pressure, density, or composition (Johnston *et al.* 1989). In general they must be obtained by fitting experimental data of a type and under conditions as close as possible to the predicted data. However, the values obtained can vary greatly with conditions, as demonstrated by the examples for methanol–CO_2 obtained by fitting the data of Brünner *et al.* (1987), and given in Table 2.2.

2.4   Intermolecular pair potentials and distribution functions for a binary mixture

In a binary mixture, molecular interactions occur between two molecules of component 1 (solvent–solvent), a molecule of component 1 with one of component 2 (solvent–solute), and two molecules of component 2 (solute–solute). There will therefore be three intermolecular pair potentials: V_{11}, V_{12} and V_{22}, respectively. V_{11} has been considered in Section 1.2.1. V_{22} will not be considered in this book in detail. In a dilute solution it will not be very significant, because solute–solute encounters will be relatively infrequent, although more important than statistics would suggest because typically solute–solute interactions are more attractive than other classes of interaction. Here we concentrate on V_{12}. This pair potential can usually only be approximately known, especially for heavy solutes, as it has to be estimated from critical parameters. Indeed the critical parameters of heavy solutes are themselves typically predictions and are imaginary in the sense that the compounds will often not be chemically stable at their predicted critical temperatures. Methods of prediction for critical parameters are given in Reid *et al.* (1987). Because of the lack of firm information, therefore, V_{12} is used only for semi-qualitative discussions of behaviour.

In terms of the Lennard–Jones pair potential (Section 1.2.1) there will also be three values of each of the parameters: σ_{11}, σ_{12}, σ_{22} and ε_{11}, ε_{12}, ε_{22}. The parameters for the solute–solute interactions can be obtained using the approximate relationships with critical parameters described in Section 1.2.1,

$$\sigma_{22} \approx 0.9(k_B T_{c,2}/p_{c,2})^{1/3} \tag{2.22}$$

$$\varepsilon_{22}/k_B \approx 0.7\,T_{c,2} \tag{2.23}$$

where $T_{c,2}$ and $p_{c,2}$ are the critical parameters of component 2. The Lennard–Jones parameters for the solvent–solute interaction are then obtained using the simple combination rules:

$$\sigma_{12} = 0.5(\sigma_{11} + \sigma_{12}) \tag{2.24}$$

$$\varepsilon_{12} = (\varepsilon_{11}\varepsilon_{22})^{1/2} \tag{2.25}$$

Using such methods, the Lennard–Jones (6 : 12) parameters for naphthalene–carbon dioxide can be calculated to be $\sigma_{12} = 0.4997$ pm and $\varepsilon_{12}/k_B = 353.4$ K. The pair potential from these parameters is shown in Fig. 2.15(b) as a function of $r^* = r/\sigma_{12}$. Also shown is a vertical dotted line at the position of the potential minimum, which occurs at $r^* = 1.122$ (see Section 1.2.1).

Similarly, in a binary mixture, there will be three pair distribution functions for the solvent–solvent, solvent–solute, and solute–solute distributions, with the notation g_{11}, g_{12}, and g_{22}, respectively. Figure 2.15(a) shows schematically some plots based on calculations of g_{12} published by Cochran and Lee (1989) from this potential at a temperature which corresponds to 308.4 K and at two densities: 1.1 times the critical density (solid line), and 1.7 times the critical density (dashed line). It can be seen that, as expected, increasing the density moves the solvent molecules around the solute molecule closer in.

2.5 Properties of binary mixtures

2.5.1 Tuning functions

The quantity $(\partial p/\partial x_i)_{T,V}$ is an important component of other thermodynamic functions, such as partial molar volumes, as shown in Section 2.3.1. It is useful in understanding processes in supercritical fluids, and so it is here given the name of tuning function. In the limit of infinite dilution, $x_2 \rightarrow 0$, it can be calculated from the van der Waals equation of state using eqn (2.11), and a calculated curve is shown in Fig. 2.16 for naphthalene in carbon dioxide at 310 K in which the parameters used were $a_{ii} = 0.267$ and 3.97 J m^3 mol^{-2} and $b_i = 31.3 \times 10^{-6}$ and 189×10^{-6} m^3 mol^{-1}, for carbon dioxide and naphthalene, respectively. These values were calculated from tabulated values of the critical pressure and temperature for naphthalene, but for carbon dioxide they

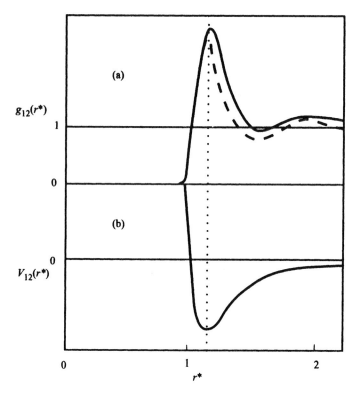

$g_{12}(r^*)$

(a)

(b)

$V_{12}(r^*)$

Fig. 2.15 (b) The Lennard–Jones (6:12) pair potential for naphthalene–carbon dioxide, V_{12}, ($\sigma_{12} = 0.4997$ pm, $\varepsilon_{12}/k_B = 353.4$ K) versus $r^* = r/\sigma_{12}$, where r is the intermolecular distance; (a) schematic pair correlation functions, g_{12}, based on calculations by Cochran and Lee (1989) from this potential at a temperature which corresponds to 308.4 K and at two densities: 1.1 times the critical density (solid line); and 1.7 times the critical density (dashed line).

were calculated from the critical density and temperature so that plots of the results are more realistically related to the critical density. At zero density, the tuning function is zero as both components are behaving as perfect gases and replacement of one by the other has no effect on the pressure. It then becomes negative, reaches a minimum and then rises in value through zero to positive values. The minimum occurs at about 540 kg m^{-3}, somewhat above the critical density of 465 kg m^{-3}. When it is negative the attractive 'a' term on the right-hand side of eqn (2.11) predominates in the calculation and when it is positive the repulsive 'b' term predominates. This indicates that the tuning function arises from attractive forces at low density and repulsive forces at high density. Figure 2.15 shows how an increase in density reduces the average distance between the solute molecule and its nearest-neighbour solvent

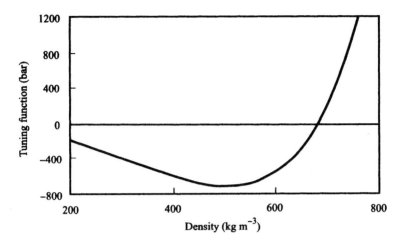

Fig. 2.16 The tuning function, $(\partial p/\partial x_2)_{T,V}$, for naphthalene in carbon dioxide at infinite dilution plotted against density, calculated from the van der Waals equation using the parameters detailed in the text.

molecules and causes the interaction to move from attractive to repulsive. This argument is given in more detail in the next section in terms of the virial theorem of Clausius. In this section, the following conclusions are given.

In a compressible medium such as a supercritical fluid it may often be possible, by controlling the density, to vary the mean distance between a solute species and its nearest-neighbour solvent molecules within a range where their mutual pair potential energy is changing significantly. The solvation energy of the solute can therefore be controlled by the density of the medium and this will cause variations in a physical or chemical equilibrium or, if the species is a transition state, the rate of a chemical reaction. The effect described by these two statements can be succinctly described as 'potential tuning' and the tuning function quantifies the process. This potential tuning is not particularly associated with the region around a critical point or line or with the critical density.

2.5.2 Discussion of tuning functions based on the virial theorem

The most physically transparent investigation of tuning functions can be made using the virial theorem (Rowlinson and Swinton 1982), which as used below assumes spherically symmetric and pair-wise additive interactions between molecules. In its original form it gives the pressure of a fluid in terms of an ensemble average, $\langle .... \rangle$, of the sum over all pairs of molecules of the intermolecular distance, r_{ij}, multiplied by the intermolecular force, f_{ij}, (defined as positive for a repulsive force)

$$p = \frac{RT}{V} + \frac{1}{3V} \left\langle \sum_i \sum_{j<i} r_{ij} f_{ij} \right\rangle \tag{2.26}$$

Thus the pressure is given as the pressure of a perfect gas corrected by a term involving intermolecular forces, which will increase the predicted pressure when the forces are repulsive and vice versa. This equation, for a single substance can be also written in terms of the pair distribution as

$$p = \frac{RT}{V} - \frac{4\pi N_A^2}{6V^2} \left(\int_0^\infty r^3 [\mathrm{d}V(r)/\mathrm{d}r] g(r) \, \mathrm{d}r \right) \tag{2.27}$$

For a binary mixture at infinite dilution, the two types of pair distributions are included and so the equation becomes

$$p = \frac{RT}{V} - \frac{4\pi N_A^2}{6V^2} \left(x_1^2 \int_0^\infty r^3 [\mathrm{d}V_{11}/\mathrm{d}r] g_{11} \, \mathrm{d}r + 2x_1 x_2 \int_0^\infty r^3 [\mathrm{d}V_{12}/\mathrm{d}r] g_{12} \, \mathrm{d}r \right) \tag{2.28}$$

The term in x_2^2, which involves g_{22}, has been omitted because of the assumed dilute conditions. On differentiation to obtain the tuning function, neglecting terms in x_2 and setting x_1 equal to unity, the following equation is obtained:

$$\left(\frac{\partial p}{\partial x_2} \right)_{T,V} = \frac{4\pi N_A^2}{6V^2} \left(\int_0^\infty r^3 \left(\frac{\mathrm{d}V_{11}}{\mathrm{d}r} \right) g_{11} \, \mathrm{d}r \right) - \frac{2\pi N_A^2}{3V^2}$$
$$\times \left(\int_0^\infty r^3 \left(\frac{\mathrm{d}V_{11}}{\mathrm{d}r} \right) \left(\frac{\partial g_{11}}{\partial x_2} \right)_{T,V} \mathrm{d}r \right)$$
$$- \frac{4\pi N_A^2}{3V^2} \left(\int_0^\infty r^3 \left(\frac{\mathrm{d}V_{12}}{\mathrm{d}r} \right) g_{12} \, \mathrm{d}r \right) \tag{2.29}$$

The first term on the right-hand side of eqn (2.29) arises because the introduction of solute causes the solvent molecules to interact with the solute and therefore there is less interaction between the solvent molecules themselves. It will be smaller than the third term because the forces between solvent molecules will be less than the solvent–solute forces. The second term is the rate at which the introduction of solute molecules disrupts the solvent structure and is difficult to quantify. However, for the sake of discussion, it will be assumed that the first and second terms are both smaller than the third term on the basis that the solute–solvent forces are stronger than the solvent–solvent forces. The major effects can then be considered using the following approximate equation:

$$\left(\frac{\partial p}{\partial x_2}\right)_{T,V} \approx -\frac{4\pi N_A^2}{3V^2} \int_0^\infty r^3 \left(\frac{dV_{12}}{dr}\right) g_{12}\, dr \qquad (2.30)$$

Equation (2.30) can now be understood in terms of Fig. 2.15, which shows both V_{12} and g_{12} at the two densities: 1.1 times the critical density (solid line); and 1.7 times the critical density (dashed line). Both are shown as functions of $r^* = r/\sigma_{12}$, and also shown is a vertical dotted line at the position of the potential minimum, to the left of which the intermolecular forces are repulsive and vice versa. As the density is increased, the first solvent shell moves in closer to the solute molecule and a higher proportion of the pair interactions will be repulsive. The integrand in eqn (2.30) will change so that at short distances the negative region will become more important. This will cause the value of the integral to fall and, therefore, the value of the tuning function to rise. The variation in the value of the tuning function can thus be seen to be an effect of the positioning of solvent molecules with respect to the solute–solvent intermolecular potential.

2.5.3 Partial molar volumes

Partial molar volumes are now calculated for the example of naphthalene in carbon dioxide in the limit of infinite dilution, $x_2 \to 0$, from the van der Waals equation of state using eqns (2.7), (2.10), and (2.11). They are shown in Fig. 2.17 for temperatures that are 1 K, 3 K, and 9 K above the critical temperature of CO_2. The parameters used, as before for the tuning function, were $a_{ii} = 0.267$ and $3.97\, J\, m^3\, mol^{-2}$ and $b_i = 31.3 \times 10^{-6}$ and $189 \times 10^{-6}\, m^3\, mol^{-1}$, for carbon dioxide and naphthalene, respectively. As before, these values were calculated from tabulated values of the critical pressure and temperature for naphthalene, but for carbon dioxide they were calculated from the critical density and temperature so that plots of the results are more realistically related to the critical density. These predictions are consistent with experimental values obtained by van Wasen and Schneider (1980).

Pure naphthalene has a molar volume of approximately $+0.15\, dm^3\, mol^{-1}$ under ambient conditions, and its partial molar volume in carbon dioxide is close to this figure at high densities. However, at lower densities it is negative and under some conditions it can be very large and negative. For example, as can be seen from Fig. 2.17, at 1 K above the critical temperature and close to the critical density, its partial molar volume is predicted by the van der Waals equation to be $-33.5\, dm^3\, mol^{-1}$. Thus if one mole of naphthalene is added to a very large amount of carbon dioxide under these conditions, the volume of the system would decrease by $33.5\, dm^3$. This spectacular behaviour is common for all solutes and for related properties.

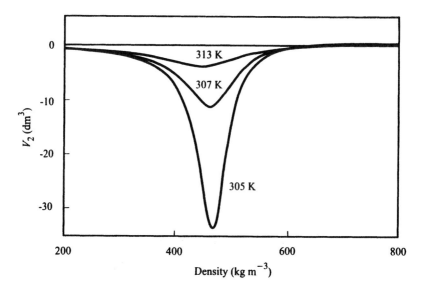

Fig. 2.17   Partial molar volumes at infinite dilution for naphthalene in carbon dioxide at three temperatures plotted against density, calculated from the van der Waals equation using the parameters detailed in the text.

2.5.4   Clustering and fluctuations in binary systems

The negative partial molar volumes and the consequent reduction in volume, described in the last section, occur because of the tendency of solvent molecules to cluster around solute molecules under this condition. The number of moles of carbon dioxide contained in the 33.5-dm^3 reduction in volume caused by one mole of naphthalene is about 400 at 305 K and the critical density. Thus 400 molecules of carbon dioxide cluster around one molecule of naphthalene under these conditions.

The theory of fluctuations, introduced in Section 1.2.3, can be extended to binary mixtures. Again a large system is considered to be divided into a large number of small systems of equal volume, v, so that molecules can pass between the systems. Now let a particular system contain N_1 and N_2 molecules of the two components. The ensemble averages $\langle N_1 \rangle$, $\langle N_2 \rangle$, and $\langle N_1 N_2 \rangle$ are now considered. The quantity $\langle N_1 N_2 \rangle - \langle N_1 \rangle \langle N_2 \rangle$ gives the tendency of the two components to associate together more than the random average, if positive, or to separate, if negative. The methods of statistical mechanics (Kirkwood and Buff 1951) can be used to show that

$$\frac{\langle N_1 N_2 \rangle - \langle N_1 \rangle \langle N_2 \rangle}{\langle N_1 \rangle \langle N_2 \rangle} = \frac{1}{N_A v}(RT\kappa_T - V_2) \qquad (2.31)$$

Near the critical point fluctuations are large in general because κ_T is large. This effect is, however, enhanced in eqn (2.31) because $-V_2$ is also large. In fact for naphthalene in CO_2, at 305 K and a density of 460 kg m^{-3}, $-V_2$ is 33.5 dm^3, and greater than $RT\kappa_T$ at 12.5 dm^3. Thus the positions of naphthalene and carbon dioxide molecules are highly correlated under these conditions, which is another indication of clustering.

Just as with a single substance, as shown in Section 1.2.3, there is a relationship between fluctuations and the pair distribution for the two components in a binary mixture. Replacing eqn (1.7) is the following relationship (Kirkwood and Buff 1951)

$$\frac{\langle N_1 N_2 \rangle - \langle N_1 \rangle \langle N_2 \rangle}{\langle N_1 \rangle \langle N_2 \rangle} = \frac{4\pi}{\nu} \int_0^\infty r^2 \left\{ g_{12} - 1 \right\} dr \tag{2.32}$$

The long tail in g_{12} for conditions near the critical point, shown in Fig. 2.15(a) (solid line) will enhance the value of the integral in eqn (2.32) and corresponds to large fluctuations in density in the critical region. As will be appreciated from the discussion in this section, there are a number of equivalent ways of describing clustering in a supercritical fluid.

3 *Solubilities of involatile substances*

3.1 Introduction

This chapter considers the behaviour and prediction of the solubility of compounds of higher molar mass, often solids, in supercritical fluids. The solubility of relatively large molecules in supercritical fluids is an important topic, which has been the subject of a large number of experimental and theoretical studies. Good data on solubilities are important in designing supercritical fluid processes, particularly supercritical fluid fractionation, as described in Chapter 6. This topic is a constituent part of the subject of Chapter 2, but has been separated because of its importance. It is difficult to make good predictions of solubilities in supercritical fluids from molecular properties in the absence of any experimental data (Johnston *et al.* 1989), although some success has been achieved for non-polar molecules in some cases and theoretical efforts are continuing. The basic problem is that there is a logarithmic relationship between the solubility and the energies in the systems, so that errors in prediction of the latter have a large effect on solubility predictions.

In this chapter, practical methods of dealing with solubility will be concentrated on, beginning with simple empirical relations. Later, the use of equations of state will be described, with the Peng–Robinson equation as an example. Finally solubilities at low pressures will be treated with the virial equation of state. The chapter begins, however, with a discussion of the general behaviour of solubility as a function of temperature and pressure.

3.2 The behaviour of solubility

The behaviour of the solubility of a heavy solute as a function of pressure at constant temperature is illustrated schematically in Fig. 3.1, in which the solid line gives the curve at lower pressures and the dashed lines show alternative behaviour at higher pressures. These curves are the portions of the curves shown in Figs 2.10, 2.11, and 2.13 for high values of x_1, rotated through 90°. The curves have the following features. There is an initial fall at low pressures, AB, which starts at $x_2 = 1$ and the solute vapour pressure, when no solvent is

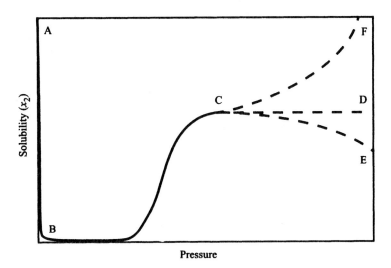

Fig. 3.1   The behaviour of solubility in a supercritical fluid with pressure at constant temperature.

present. As the solvent is added and the solute is diluted without being much solvated, x_2 falls towards B. This initial fall is at very low pressures for solutes of low volatility, and in many cases will be virtually coincident with $p = 0$. The second feature is a rise in solubility BC, which is a feature of all solubility data. The rise is due to solvation arising from attractive forces between the solvent and solute molecules. It has a steep portion near the critical pressure of the solvent, where its density is rising most rapidly. Thereafter the solubility may be approximately constant, represented by CD. Alternatively, there may be a fall as in CE. If this occurs, it is because at higher pressures, where the solvent is becoming compressed and repulsive solute–solvent interactions are becoming important, the solute chemical potential is raised to a greater extent than in the solute-rich liquid or solid phase. Finally, a rise, CF, may occur if there is a critical line present at high pressures at the temperature of the isotherm, whereupon the solubility will rise towards it as shown also in Figs 2.11 and 2.13(b).

At constant pressure, solubility varies with temperature as shown in Fig. 3.2, with the solubility initially falling, reaching a minimum, and then rising again. The full range of behaviour, however, is not shown by all compounds under all conditions. It will be shown later that solubility in a supercritical fluid depends on two main factors: the vapour pressure of the solute and the solvating effect of the fluid, which depends on density. The fall at lower temperatures occurs when the density falls, reducing the solvating effect to a greater extent than is compensated for by the rise in vapour pressure.

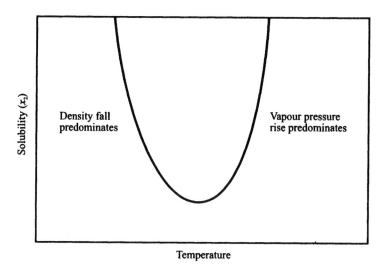

Fig. 3.2   The behaviour of solubility in a supercritical fluid with temperature at constant pressure.

Eventually, however, the vapour pressure rise, which is approximately exponentially dependent on absolute temperature, takes over, reversing the trend. The minimum in solubility occurs at different temperatures for different compounds, which sometimes facilitates separations, for example in recrystallization (Caralp *et al.* 1993). The temperature of the minimum also changes with pressure for the same compound, usually moving to higher temperatures for lower pressures, where the density change is more dramatic. Consequently, within a given temperature range, the solubility may be falling with temperature at lower pressures, but rising with temperature at higher pressures. This can cause the isotherms of solubility versus pressure to cross, as shown in Fig. 3.3 for naphthoquinone in carbon dioxide (Schmidt and Reid 1986), giving rise to the so-called cross-over effect. The effect can be seen more clearly in Fig. 3.7 below from predicted curves.

3.3   Empirical correlation of solubilities

The first approach described to rationalize experimental solubility data is to fit them to simple empirical equations. The data can then be predicted for intermediate conditions and, to a limited extent, extrapolated outside the experimental range. The equations can also be easily inserted into design calculations. The empirical approach described here is based on the assumption that the solubility of a heavy solute in a supercritical fluid arises from the vapour pressure of the solute, which is dependent on temperature, being enhanced by the solvating effect of the fluid, which is dependent on density.

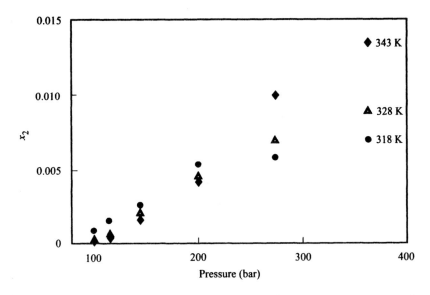

Fig. 3.3   Isotherms of the solubility of naphthoquinone in carbon dioxide, showing the cross-over effect (Schmidt and Reid 1986).

This leads to the definition of an *enhancement factor*, E (Johnston *et al.* 1989) as the ratio of the partial pressure of the solute in the supercritical phase to the vapour pressure of the pure solute at the same temperature, p_v,

$$E = x_2 p / p_v \qquad (3.1)$$

The natural logarithm of the enhancement factor is then fitted to a linear function of density, i.e.

$$\ln E = A + c\rho \qquad (3.2)$$

where A and c are constants. Such linear functions of density have also been used for simpler fits of the logarithm of mole fraction itself, or solubility expressed as concentration, when the constants become functions of temperature. A study of the correlation of solubilities of some 90 compounds showed that the approach here is somewhat, but not greatly, superior to these simpler correlations (Bartle *et al.* 1991a).

A practical problem in making these correlations is that the constant A, the intercept at zero density, is extrapolated a long way from the experimental data. Values therefore are subject to large errors. To avoid this, eqn (3.2) is modified to introduce a reference density, close to the experimental densities:

$$\ln E = A' + c(\rho - \rho_{ref}) \qquad (3.3)$$

A value of ρ_{ref} of $700\,kg\,m^{-3}$ is typically suitable and the new constant A', which is the intercept at ρ_{ref}, is given by $A + c\rho_{ref}$. A further problem is that the vapour pressures, needed to calculate E, are not known or well known for many of the solutes, particularly the solids. Substituting eqn (3.1) into eqn (3.3) and rearranging gives

$$\ln(x_2 p/p_{ref}) = \ln(p_v/p_{ref}) + A' + c(\rho - \rho_{ref}) \tag{3.4}$$

or

$$\ln(x_2 p/p_{ref}) = A'' + c(\rho - \rho_{ref}) \tag{3.5}$$

where $A'' = \ln(p_v/p_{ref}) + A'$ and p_{ref} is a reference pressure, for which 1 bar (or 1 atmosphere) is a suitable value. A'' can be assumed to have the following dependence on temperature:

$$A'' = a + b/T \tag{3.6}$$

based on the approximate behaviour of vapour pressure with temperature, where a and b are constants.

The conformation of real experimental data to the behaviour of the above equations is now examined, using results for naphthalene, which has been studied by several research groups. Table 3.1 gives values for c and A'' obtained by fitting 16 isotherms from seven published sources individually to eqn (3.5) using $p_{ref} = 1$ bar and ρ_{ref} $700\,kg\,m^{-3}$. As can be seen c, whilst showing some variation due partly to experimental error, shows no tendency to change over the experimental range. The conformation of A'' to eqn (3.6) is examined in Fig. 3.4 and again shows satisfactory behaviour. The slope of this line corresponds to an enthalpy change of $77\,kJ\,mol^{-1}$, compared with an accepted value

Table 3.1 Parameters obtained by fitting isotherms from seven sources for the solubility of naphthalene in carbon dioxide (Bartle et al. 1991a)	$T(K)$	$10^3 \times c\ (m^3\,kg^{-1})$	A''
	308	8.00	−0.1394
	308	7.82	−0.1121
	308	8.50	−0.1517
	308	8.86	−0.2659
	308	8.45	−0.1700
	318	6.50	0.8542
	318	6.07	0.8673
	323	6.70	1.2267
	328	7.50	1.7216
	328	7.84	1.6802
	328	7.59	1.6435
	328	7.91	1.7143
	328	7.41	1.7125
	331	8.02	1.8086
	333	8.50	2.3333
	338	8.77	2.5105

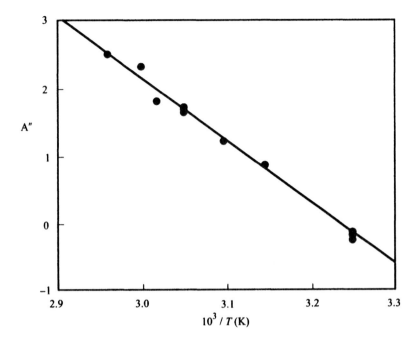

Fig. 3.4 Plot of the empirical parameter, A'', versus reciprocal absolute temperature.

for the enthalpy of vaporization of solid naphthalene of $70 \, \text{kJ mol}^{-1}$. This shows that the temperature dependence of A'' is mainly due to the variation of vapour pressure, as is consistent with the above arguments. From the behaviour of the naphthalene data, it can therefore be concluded that the above equations form a good basis for an empirical correlation.

Combining eqns (3.4) and (3.5) gives the final correlation equation

$$\ln(x_2 p/p_{\text{ref}}) = a + b/T + c(\rho - \rho_{\text{ref}}) \qquad (3.7)$$

Measured solubility data for the dye Disperse Yellow 82 (3-(1*H*-benzo-imidazol-2-yl)-7-(diethylamino)chromen-2-one) at $353 \, \text{K}$ and $373 \, \text{K}$ were fitted to eqn (3.7) to give (Özcan *et al.* 1997)

$$\ln(x_2 p/\text{bar}) = 21.12 - 101\,991/T/\text{K} + 0.0116(\rho/\text{kg m}^{-3} - 700) \qquad (3.8)$$

Figure 3.5 shows the experimental points and the theoretical curves (solid lines) at the two temperatures. The fit is good over the small range of conditions. The fit for data on five polynuclear hydrocarbons over a temperature range from $313 \, \text{K}$ to $523 \, \text{K}$ is worse, but still acceptable (Miller *et al.* 1996).

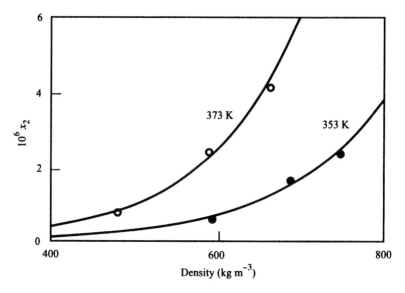

Fig. 3.5.   Measured solubility data for the dye Disperse Yellow 82 (3-(1*H*-benzoimidazol-2-yl)-7-(diethylamino)chromen-2-one) at 353 K (filled points) and 373 K (open points) (Özcan *et al.* 1997), with predictions from the curve $\ln(x_2 p/\text{bar}) = 21.12 - 101\,991/T/\text{K} + 0.0116(\rho/\text{kg m}^{-3} - 700)$ shown as solid curves.

3.4   Calculation of solubilities from the Peng–Robinson equation of state

A more sophisticated method of treating solubilities is to use an equation of state. Again the Peng–Robinson equation is given as an example, as it is the equation most used in supercritical fluids. The equation can be used to predict phase behaviour for solutes with a range of volatility, as described in Chapter 2. Here it is simplified for only slightly volatile, solid solutes. As described earlier, the equation has an unknown interaction parameter, k_{12}, for a binary mixture, to which the predicted solubility is sensitive. For this reason, although k_{12} can be obtained in principle from many types of physical property data for the mixture, it usually needs to be obtained by fitting experimental solubilities to the equation. Further difficulties are that the interaction parameters are often found to be temperature dependent, the equation does not fit the data equally well at all temperatures and pressures, and reliable values of the other physical parameters needed for the equation of state are not always available.

It is assumed in what follows that the solute in equilibrium with the saturated supercritical solution is a solid and that it contains a negligible amount of the supercritical fluid substance (component 1). It is also assumed that the partial molar volume of the solid solute in the system, V_2, at all pressures and temperatures considered, is equal to its molar volume, V_m, at atmospheric pressure and 298 K, which will be close to the truth. The thermodynamic formulae for the equilibrium are established by equating the fugacities of component 2 in the solid and supercritical fluid phases. The standard state for the fugacities will be taken throughout as that appropriate for the gas and supercritical fluid phases, i.e. that $f_2 = px_2$ in the limit of zero pressure. As solutes of low volatility are being considered, it will be assumed that the vapour of the pure solvent is behaving as a perfect gas and that, therefore, the fugacity of the solid at a particular temperature and at a pressure close to zero will be given by $p_v(T)$, the vapour pressure of component 2 at T. Given the standard relationship that $RT[\partial \ln(f_2/p^{\ominus})/\partial p]_T = V_2$, and that V_2 is assumed to be always equal to V_m, the fugacity of the solid at pressure p will be given by

$$\ln(f_2/p^{\ominus}) = \ln(p_v/p^{\ominus}) + pV_m/RT \qquad (3.9)$$

At equilibrium, this will be equal to the fugacity of component 2 in the fluid phase, which is equal to $px_2\phi_2/p^{\ominus}$. The saturation value of x_2, the solubility, will therefore be given by

$$\ln x_2 = \ln(p_v(T)/p) - \ln \phi_2 + pV_m/RT \qquad (3.10)$$

For the Peng–Robinson equation, ϕ_2 for a dilute solution is given by eqn (2.21).

The equations for calculating solubility under the circumstances given above are given in Table 3.2. These equations allow the calculation of solubility at a given temperature and solvent molar volume, given the vapour pressure and molar volume of the solute, the critical parameters and acentric factors of both components, and the binary interaction parameter. The pressure is also calculated during this procedure, and also the density, ρ, can be assumed, at

Table 3.2   Simplified Peng–Robinson equations for the mole fraction, x_2, at saturation of a solute of low volatility forming a dilute solution in a supercritical fluid

$$\ln x_2 = \ln(p_v(T)/p) - \ln \phi_2 + pV_m/RT$$

$$\ln \phi_2 = \ln\left(\frac{RT}{p(V-b_1)}\right) + \frac{b_2}{b_1}\left(\frac{pV}{RT}-1\right) - \frac{a_{11}}{2\sqrt{2}RTb_1}\left(\frac{2a_{12}}{a_{11}}-\frac{b_2}{b_1}\right)\ln\left(\frac{V+(1+\sqrt{2})b_1}{V+(1-\sqrt{2})b_1}\right)$$

$$p = \frac{RT}{V-b_1} - \frac{a_{11}(T)}{V^2+2Vb_1-b_1^2}$$

$$b_1 = 0.0778RT_{c,1}/p_{c,1}; \qquad b_2 = 0.0778RT_{c,2}/p_{c,2}$$

$$a_{12} = (0.45724(1-k_{12})R^2T_{c,1}T_{c,2}\kappa_1\kappa_2)/\sqrt{p_{c,1}p_{c,2}}; \quad a_{11} = (0.45724R^2T_{c,1}^2\kappa_1^2)/p_{c,1}$$

$$\kappa_1 = 1 + (0.37464 + 1.54226\omega_1 - 0.26992\omega_1^2)(1-\sqrt{T/T_{c,1}})$$

$$\kappa_2 = 1 + (0.37464 + 1.54226\omega_2 - 0.26992\omega_2^2)(1-\sqrt{T/T_{c,2}})$$

T is the absolute temperature; V the molar volume of the pure solvent; $p_v(T)$ the vapour pressure of the solute; V_m the molar volume of the pure solute; $p_{c,i}$, $T_{c,i}$, and ω_i are critical pressures, temperatures, and acentric factors of the solvent (1) and the solute (2), and k_{12} is the binary interaction parameter.

dilute conditions, to be equal to that of the pure fluid, i.e. $\rho = M_1/V$, where M_1 is the molar mass of component 1. The final outcome is a value of solubility for a particular temperature and pressure or density. Tables 3.3 and 3.4 give the necessary parameters for some examples of solutes.

Table 3.3   Critical parameters, acentric factors, ω, and binary interaction parameters with carbon dioxide, k_{12}, for some solutes from a number of sources (Bartle et al. 1992)

Solute	T_c (K)	p_c (bar)	ω	k_{12}
n-Docosane	787	9.93	0.976	0.128
n-Tetracosane	806	9	1.066	0.140
1-Octadecanol	790	14	0.892	0.056
Stearic acid	810	17	1.085	0.071
Palmitic acid	791	19	1.047	0.083
Oleic acid	797	17	1.120	0.103
Naphthalene	750	41	0.302	0.085
Phenanthrene	883	32	0.437	0.125
Anthracene	869	22	0.370	0.090
Fluorene	826	30	0.406	0.090
Pyrene	895	26	0.465	0.092
Biphenyl	769	34	0.416	0.095
Carbazole	899	33	0.496	0.193
Acridine	890	30	0.428	0.126
Phenol	692	61	0.450	0.104
p-Chlorophenol	730	48	0.490	0.098
2-Naphthol	834	42	0.480	0.284
Benzoic acid	752	46	0.620	0.127

Table 3.4   Parameters for solute vapour pressures, p_v, given by $\log_{10}(p_v/\text{bar}) = -c/T + d$, and values for the solute molar volume V_m from a number of sources (Bartle et al. 1992)

Solute	c (K)	d	$10^6 V_m$ (m³)
n-Docosane	6258	13.79	391
n-Tetracosane	7159	15.40	442
1-Octadecanol	11589	27.45	333
Stearic acid	9509	20.85	302
Palmitic acid	8376	18.41	301
Oleic acid	6497	12.39	316
Naphthalene	3729	8.57	125
Phenanthrene	4500	8.42	182
Anthracene	5290	9.72	139
Fluorene	2095	8.32	138
Pyrene	4950	8.52	159
Biphenyl	4330	9.62	178
Carbazole	4836	8.25	152
Acridine	4826	9.30	178
Phenol	2598	5.63	89
p-Chlorophenol	3572	8.25	102
2-Naphthol	3089	5.51	185
Benzoic acid	3333	6.15	96

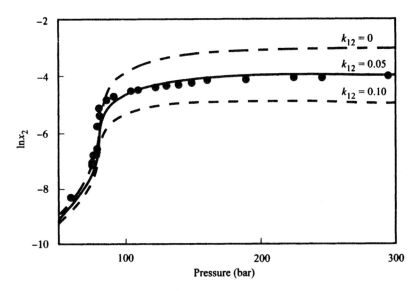

Fig. 3.6 Curves showing predictions of the solubility of naphthalene in carbon dioxide at 308 K for the Peng–Robinson equation of state using three different binary interaction parameters, compared with the experimental data of Tsekhanskaya *et al.* (1964) (filled points).

Figure 3.6 shows some predictions of solubility for naphthalene in carbon dioxide using different binary interaction parameters, which shows how sensitive the values are to this parameter, with x_2 changing by a factor of about 2.5 at higher pressures when k_{12} changes by 0.05 (Bartle *et al.* 1990a). Also shown are some of the earliest experimental data for comparison (Tsekhanskaya *et al.* 1964), and the figure therefore also illustrates how the experimental data determine the binary interaction parameter. If isotherms are fitted over a range of temperatures for naphthalene in carbon dioxide, using data from a number of sources, the binary interaction parameters obtained show a strong dependence on temperature, as shown in Table 3.5. This is probably because the higher temperatures are approaching that where naphthalene can form a liquid phase in the presence of carbon dioxide under pressure (340 K), as described in Section 2.2.3, and so the assumptions made for the present calculation are not valid. The values fall from 0.085 to 0.013 as the temperature rises from 308 K to 338 K.

Similar behaviour is shown by phenol, and also, to a smaller extent over the same temperature range, by octadecanol, biphenyl, and benzoic, stearic, and oleic acids. Other compounds give interaction parameters which are constant within the probable errors over this same temperature range. An example is phenanthrene, for which the data are also given in Table 3.5, and are seen to be all within 0.01 of a value of 0.12. This approximately constant behaviour is

	T (K)	k_{12}
Naphthalene	308	0.085
	318	0.075
	323	0.072
	328	0.052
	332	0.055
	333	0.014
	338	0.013
Phenanthrene	308	0.126
	313	0.123
	318	0.117
	323	0.122
	328	0.111
	338	0.107
	343	0.127

Table 3.5   Values of the Peng–Robinson interaction parameter, k_{12}, for naphthalene or phenanthrene and carbon dioxide at various temperatures, obtained by fitting experimental solubility isotherms (Bartle *et al.* 1990a)

also exhibited by fluorene, pyrene, anthracene, and acridine. Thus for most compounds, which have melting points well above the temperature range of interest, it can be assumed that the binary interaction parameters do not show a marked trend with temperature.

The binary interaction parameters given in Table 3.3 were obtained by fitting solubility isotherms close to 308 K. These values can be used to predict solubilities, using the equations of Table 3.2, for most of the compounds over the whole temperature range. Figure 3.7 shows an example of predictions for 2-naphthol in carbon dioxide, which exhibit the cross-over effect, described in Section 3.2.

3.4.1   Estimation of binary interaction parameters for the Peng–Robinson equation of state

If there are no experimental solubility data, or other suitable data, it is desirable to be able to guess the binary interaction parameter so that at least some estimate of solubility may be obtained using the equations of Table 3.2. With the exception of carbazole, all the compounds in Table 3.3 have k_{12} values between 0.05 and 0.15. Unsophisticated methods would be to use the value for a similar compound or an average value of 0.10. A slightly improved method is to use a published correlation (Bartle *et al.* 1992) to predict k_{12} values, which gives the formula

$$k_{12} = 0.51B; \qquad B = A(\omega_2 - \omega_1)(V_{c,2}/V_{c,1})(p_{c,2}/p_{c,1})^2 \qquad (3.11)$$

where A is a parameter which is $1/2$ for compounds containing –OH groups and 1 for other compounds. Critical volumes, where not available, can be calculated from (Reid *et al.* 1987)

$$V_{c,i} = (0.2918 - 0.928\omega_i)RT_{c,i}/p_{c,i} \qquad (3.12)$$

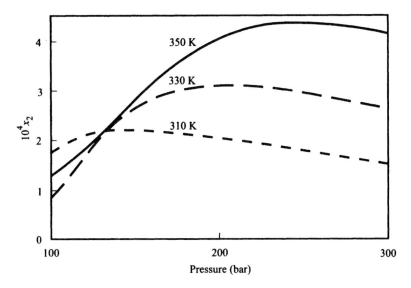

Fig. 3.7   Three isotherms for the solubility of 2-naphthol in carbon dioxide predicted from the Peng–Robinson equation of state, showing the cross-over effect.

Figure 3.8 shows this correlation compared with binary interactions for 19 compounds. The root-mean-squared deviation from the line vertically is 0.015, which is a typical probable error in the calculated values of k_{12}. However, some individual points deviate from the line well outside their calculated probable errors, the worst case being naphthalene, with a deviation of 0.031. Solubilities for naphthalene at 308 K, calculated using a value of k_{12} obtained from eqn (3.11), the simplified Peng–Robinson equations of Table 3.2, and the parameters of Tables 3.3 and 3.4, are higher than experimental solubilities by around 50 per cent.

3.5   Calculation of solubilities at low pressures using the virial equation

It is sometimes important to estimate solubilities at low pressure, because these are important when components in a process are separated by reducing the fluid pressure. Not all the component will be separated, some being lost in effluent streams or recycled with the solvent. In addition further material will be entrained as an aerosol at the end of a separation. Under these circumstances, the virial equation of state may be useful and will often give reasonable estimates at densities up to about one half of the critical density of the

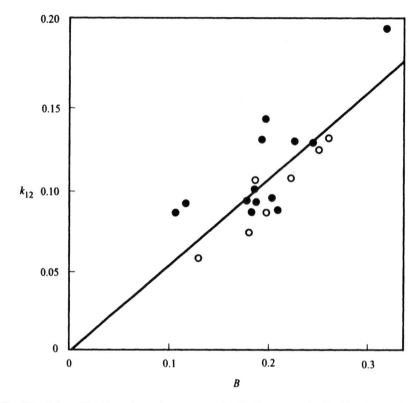

Fig. 3.8 Values of the binary interaction parameter, k_{12}, for 19 compounds with CO_2, obtained from experimental solubility data, plotted versus $B = A(\omega_2 - \omega_1)(V_{c,2}/V_{c,1})(p_{c,2}/p_{c,1})^2$, for compounds without (solid points) and with OH groups (open points), and compared with the prediction that $k_{12} = 0.51B$ (solid line).

solvent. The virial equation of state was introduced in Section 1.3.1 and in this section we use the dilute-gas approximation, which applies when only the second virial coefficient, $B(T)$, is of numerical significance. It can be shown theoretically that the second virial coefficient for a binary mixture is given by

$$B(T) = x_1^2 B_{11}(T) + 2x_1 x_2 B_{12}(T) + x_2^2 B_{22}(T) \qquad (3.13)$$

where the B_{ij} arise from intermolecular interactions between species of components i and j and can be calculated from the intermolecular pair potential function using equations analogous to those given in Section 1.3.1. The pair potentials for the interactions involving solutes can be estimated using the methods given in Section 2.4. In the dilute-gas approximation, the virial equation of state for a binary mixture becomes

$$\frac{p}{RT} = \frac{1}{V} + \frac{1}{V^2} \left(x_1^2 B_{11}(T) + 2x_1 x_2 B_{12}(T) + x_2^2 B_{22}(T) \right) \tag{3.14}$$

An expression for the fugacity must be sought from the virial equation of state so that it can be inserted into eqn (3.10) for the solubility. This can be obtained using the methods and equations detailed for the van der Waals equation and given in Section 2.3.1 and is

$$\ln \phi_2 = \frac{p}{RT} (2B_{12} - B_{11}) \tag{3.15}$$

The equations needed for a calculation of solubility using the virial equation of state are collected together in Table 3.6 for convenience. These are now used to calculate the solubility of naphthalene in carbon dioxide at low pressure. Using the Lennard–Jones parameters for CO_2 given in Table 1.2 and those for the naphthalene–CO_2 interaction given in Section 2.4, values of B_{11} and B_{12} of -0.00011 and $-0.00051 \, m^3 \, mol^{-1}$ were obtained, respectively, at 308 K. The solubility curve obtained from these up to 50 bar is shown as the continuous line in Fig. 3.9. Numerical values of solubility measurements made at low pressures were not given (Najour and King 1966), but the experimental measurements were used to determine second virial coefficients. A value for naphthalene–CO_2 of $-0.00050 \, m^3 \, mol^{-1}$ at 309 K is

---

Table 3.6 **Equations for the mole fraction, x_2, at saturation at low pressure of a solute forming a dilute solution in a supercritical fluid derived from the virial equation of state**

$$\ln x_2 = \ln(p_v(T)p) - \ln \phi_2 + pV_m/RT; \qquad \frac{p}{RT} = \frac{1}{V} + \frac{B_{11}}{V^2}; \qquad \ln \phi_2 = \frac{p}{RT}(2B_{12} - B_{11})$$

$$B_{ij} = 2\pi N_A \sigma_{ij}^3 B_{ij}^*/3; \qquad B_{ij}^* = -\frac{\sqrt{2}}{4} \sum_{j=0}^{\infty} \left[\left(\frac{1}{T_{ij}^*} \right)^{(2j+1)/4} \frac{2^j}{j!} \Gamma \left(\frac{2^j - 1}{4} \right) \right]$$

$$\sigma_{12} = \tfrac{1}{2}(\sigma_{11} + \sigma_{22}); \qquad \sigma_{22} = 0.5(k_B T_{c.2}/p_{c.2})^{1/3}$$

$$T_{ij}^* = k_B T/\varepsilon_{ij}; \qquad \varepsilon_{12} = \sqrt{\varepsilon_{11}\varepsilon_{22}}; \qquad \varepsilon_{22} = 0.7 \, T_{c.2}$$

(B_{ij} can be readily calculated in computer programs and spreadsheets and is available in tabular form in some textbooks (e.g. Maitland *et al.* 1981).

ε_{11} and σ_{11} values for common substances are tabulated in a number of publications and some are given in Table 1.2.

T is the absolute temperature, σ_{11}, ε_{11}, and V the Lennard–Jones parameters and molar volume of the pure solvent, respectively, and $p_v(T)$, V_m, $p_{c,2}$, and $T_{c,2}$ the vapour pressure, molar volume, and critical parameters of the pure solute, respectively.

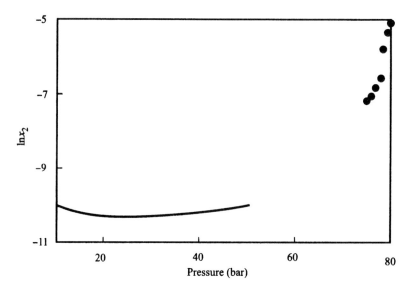

Fig. 3.9   The solubility of naphthalene in carbon dioxide at 308 K and low pressures, showing the minimum in solubility expressed as mole fraction, and given predictions from the virial equation of state (solid line) and the experimental data of Tsekhanskaya *et al.* (1964) (filled points).

given, indicating that the experimental solubility measurements are consistent with the present calculation. Experimental measurements made at higher pressures by Tsekhanskaya *et al.* (1964) are also shown in Fig. 3.9. These together illustrate the low pressure behaviour, discussed earlier in Section 3.2 and shown in Fig. 3.1.

4 *Transport properties*

4.1 Introduction

The three transport properties that will be discussed in this chapter, in order of their importance in supercritical fluids, are diffusion, viscosity and thermal conductivity. Diffusion is the transport of mixture components down a concentration gradient. Viscosity is the transport of transverse momentum along a transverse velocity gradient, which requires pressure gradients to cause fluid flow. Thermal conductivity is the transport of energy down a temperature gradient. As the critical point is approached closely, these transport properties can exhibit interesting behaviour. This is a subject much researched (e.g. Sengers 1994), but will not be discussed here. However, even in supercritical fluids in general, and particularly in the region around the critical point, there is some notable behaviour, which will be discussed.

The transport properties give rise to many of the advantages that supercritical fluids may have for a particular experiment or process. Generally, diffusion is faster and viscosity is lower than in a liquid, and this arises partly from the lower densities used in supercritical fluids and partly because supercritical fluid substances are composed of relatively small mobile molecules. Table 4.1 gives some values to illustrate this. Faster diffusion can give rise to more rapid extraction, more rapid separation in chromatography and more rapid rates for heterogeneous and diffusion-controlled homogeneous chemical reactions. Lower viscosity means that pumping fluids, particularly through packed beds, is easier.

Table 4.1 The viscosity, η, (Vesovic et al. 1990) of carbon dioxide and the diffusion coefficient for naphthalene in carbon dioxide, D (Clifford and Coleby 1991) under gas, supercritical, and liquid conditions

	η (μPa s)	D (m^2 s^{-1})
Gas, 313 K, 1 bar	16	5.1×10^{-6}
Supercritical, 313 K, 100 bar	17	1.4×10^{-8}
Liquid, 300 K, 500 bar	133	8.7×10^{-9}

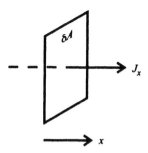

To define transport properties, the concept of a *flux* is used. Flux is a vector quantity, but for simplicity it will first be considered only in the x-direction. An elemental area δA is first defined perpendicular to the x-direction, as shown. For diffusion, molecules of, say, component 1 will be passing both ways through the area and there will be a net number passing in one direction. The net number of moles of component 1, δn_1, passing through the elemental area in the positive direction in time δt is then used to define the molar flux of component 1 in the x-direction, $J_{x,1}$, by

$$J_{x,1} = \frac{\delta n_1}{\delta A \delta t} \rightarrow \frac{\mathrm{d}^2 n_1}{\mathrm{d}A \, \mathrm{d}t} \tag{4.1}$$

with the final term being the limit as the δ quantities go to zero. Thus the flux in the x-direction at some point in the fluid and at a particular time is the net number of moles of the component moving through an area perpendicular to this direction, per unit area in unit time. The units of this flux will be $\mathrm{mol\,m^{-2}\,s^{-1}}$ in the SI system. Inherent in this discussion, is an assumption that the *frame of reference* of the coordinates is known, but this concept will be ignored in this initial discussion and treated in the next section.

Fick's first law of diffusion, which is a physical (experimental) law, states that flux of the component is proportional to its concentration gradient. Again, only concentration gradients and flux in the x-direction are considered and those in other directions are considered to be zero. The law can then be written as

$$J_{x,1} = -D_1 \frac{\mathrm{d}c_1}{\mathrm{d}x} \tag{4.2}$$

where c_1 is the concentration of component 1 in moles per unit volume and D_1 is the *diffusion coefficient* of component 1. The minus sign indicates that the net flux is in the direction of decreasing concentration. The dimensions of the diffusion coefficient are area per unit time, and the units are $\mathrm{m^2\,s^{-1}}$ in the SI system. In a fluid of three dimensions, the flux will be a vector in the direction of concentration change and eqn (4.2) becomes

$$\boldsymbol{J}_1 = -D_1 \operatorname{grad} c_1 \tag{4.3}$$

For thermal conductivity, a temperature gradient and energy flux in the x-direction only is first considered. The energy flux, $J_{x,E}$, is defined in an analogous way to that used for the mass flux of component 1, and is a net flow of energy per unit time per unit area at any particular point in space and time. Its units in the SI system are therefore $W\,m^{-2}$. A physical law also relates the energy flux and the temperature gradient, i.e.

$$J_{x,E} = -\lambda \frac{dT}{dx} \tag{4.4}$$

where λ is the *thermal conductivity*, with SI units of $W\,m^{-1}\,K^{-1}$. In three dimensions, the equation becomes

$$J_E = -\lambda\,\text{grad}\,T \tag{4.5}$$

A related quantity is the *thermal diffusivity*, $\lambda/\rho C_p$, where ρ is the density, here in moles per unit volume, and C_p the molar heat capacity at constant pressure, i.e. ρC_p is the heat capacity per unit volume at constant pressure. The units of thermal diffusivity are the same as those for the diffusion coefficient, i.e. $m^2\,s^{-1}$.

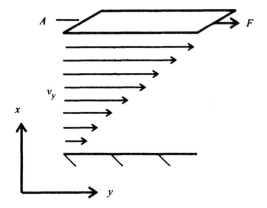

Viscosity arises from the transfer of transverse momentum and for simplicity it is considered that the fluid is moving only in the y-direction and that its velocity is only changing in the x-direction. This process is less easy to visualize and, although similar arguments and equations can be developed for viscosity as were given for diffusion and thermal conductivity, a phenomenological approach will be used here, which is equivalent, and this is illustrated in the diagram, above. A parallel-sided slab of fluid is considered which is sandwiched between a stationary base and a moving plate of area A. A force F is applied to the plate to cause it to move and set up, at steady state, fluid velocities in the y-direction, v_y, which have a gradient in the x-direction, as shown. The relationship between the force applied and the velocity gradient is found experimentally to be

$$F/A = \eta \frac{\mathrm{d}v_y}{\mathrm{d}x} \qquad (4.6)$$

where η is the dynamic viscosity coefficient, with SI units of Pa s.

A related quantity known as the kinematic viscosity coefficient, which is a ratio of the dynamic viscosity and the density in terms of mass per unit volume, η/ρ, is also used, which has SI units of $m^2 s^{-1}$. Thus diffusion coefficients, thermal diffusivities, and kinematic viscosities all have the same dimensions and units. Table 4.2 gives a summary of the transport coefficients discussed with their symbols and SI units. Finally, it should be noted that there are particular circumstances, not discussed in the present context, where non-linear responses to gradients can occur and the analysis given here is insufficient.

Equations for the transport coefficients at low pressures were first obtained from the simple kinetic theory of gases, which calculates the flux of quantities on the assumption that motion is randomized after collision. In fact there is persistence of velocity (including direction) after a molecular collision to some extent. This is taken into account in the rigorous kinetic theory developed by Chapman and Enskog at the beginning of the twentieth century, and the equations derived (Maitland *et al.* 1981) are now routinely used and require knowledge of the pair potential for the two colliding molecules. The relationships are valid, depending however on the accuracy required, below 1 bar. At low pressures, dynamic viscosities and thermal conductivities are independent of density, but diffusion coefficients, thermal diffusivities, and kinematic viscosities are inversely proportional to the density. For somewhat higher densities, the Thorne–Enskog theory (Maitland *et al.* 1981) can be used to predict transport properties. As the density becomes higher again, however, prediction is difficult.

Table 4.2   Transport properties discussed in this chapter

Transport property	Symbols	SI units
Diffusion coefficient	D, D_{12}	$m^2 s^{-1}$
Dynamic viscosity	η	Pa s
Kinematic viscosity	η/ρ	$m^2 s^{-1}$
Thermal conductivity	λ	$W m^{-1} K^{-1}$
Thermal diffusivity	$\lambda/\rho C_p$	$m^2 s^{-1}$

4.2 Diffusion

4.2.1 Diffusion in binary and multicomponent mixtures

Diffusion is more complex than has been indicated in the previous section, especially in a multicomponent mixture. Because of the sensitivity of properties in supercritical fluids to concentration, these complexities often need to be understood in the present context, whereas they can often be neglected in other situations. As previously mentioned, diffusion is considered relative to a defined frame of reference, which may be moving with respect to space-fixed coordinates. Initially a binary mixture is considered in which diffusion occurs only in the x-direction and equations of the form of eqn (4.2) can be written for both components, as follows:

$$J_{x,1} = -D_1 \frac{dc_1}{dx} \tag{4.7}$$

$$J_{x,2} = -D_2 \frac{dc_2}{dx} \tag{4.8}$$

The frame of reference for these general equations is now chosen so that the two diffusion coefficients are identical, this being the case when there is no change of volume across the perpendicular plane containing the elemental area when diffusion occurs. When this *volume-fixed reference frame* is used, it will determine the values of the fluxes, so these are now given a superscript and denoted by $J_{x,1}^v$ and $J_{x,2}^v$. The rate of change in volume that occurs, per unit area of the surface, as each of the two components flow will be equal to their fluxes multiplied by their partial molar volumes. For no net volume change, the sum of these products must be zero, i.e.

$$J_{x,1}^v V_1 + J_{x,2}^v V_2 = 0 \tag{4.9}$$

Substituting for the two fluxes using eqns (4.7) and (4.8) gives

$$D_1 V_1 \frac{dc_1}{dx} + D_2 V_1 \frac{dc_2}{dx} = 0 \tag{4.10}$$

There is a relationship between the two partial molar volumes of the Gibbs–Duhem form, which is

$$V_1 dc_1 + V_2 dc_2 = 0 \tag{4.11}$$

and comparison of equations (4.10) and (4.11) shows that the two diffusion coefficients are identical and both can now be denoted by D_{12}, which is known as the *binary diffusion coefficient*.

The volume-fixed frame of reference is convenient for diffusion in dilute solution in liquids, but is not always convenient for supercritical fluids. For this situation the *mass-fixed reference frame* or *barycentric frame* may be more suitable, where there is no total movement of mass across the plane in which the elemental area is set. The fluxes in this reference frame are now defined as the mass of the components passing per unit area in time and are denoted by $J_{x,1}^m$ and $J_{x,2}^m$. In the mass-fixed reference frame

$$J_{x,1}^m + J_{x,2}^m = 0 \tag{4.12}$$

The fluxes in terms of moles will be given by $J_{x,1}^m/M_1$ and $J_{x,2}^m/M_2$ where M_i are the molar masses of the components. The volume increase that occurs on the negative side of the reference frame, per unit area in unit time, as a results of these fluxes will have the dimensions of a velocity, v_{mv}, and will be given by

$$v_{mv} = -J_{x,1}^m V_1/M_1 - J_{x,2}^m V_2/M_2 \tag{4.13}$$

which, because of eqn (4.12) can be simplified to

$$v_{mv} = J_{x,1}^m (V_2/M_2 - V_1/M_1) \tag{4.14}$$

The velocity v_{mv} is, on consideration, the velocity of the mass-fixed reference frame with respect to the volume-fixed reference frame. The difference between the fluxes in the two reference frames can be obtained from this velocity multiplied by the concentration of the appropriate component, i.e.

$$J_{x,1}^m/M_1 - J_{x,1}^v = -v_{mv}c_1 = -J_{x,1}^m (V_2/M_2 - V_1/M_1)c_1 \tag{4.15}$$

Both terms on the left-hand side of eqn (4.15) are given in terms of moles per unit volume in unit time, but that in the mass-fixed reference frame will be lower because the frame has moved forwards passing $J_{x,1}^m (V_2/M_2 - V_1/M_1)c_1$ moles of component 1 per unit area in unit time.

Using the thermodynamic relationship, which arises directly from the definition of partial molar volumes, that

$$c_1 V_1 + c_2 V_2 = 1 \tag{4.16}$$

eqn (4.16) can be simplified to

$$J_{x,1}^v = J_{x,1}^m (V_2/M_1 M_2)(M_1 c_1 + M_2 c_2) \tag{4.17}$$

As $M_1 c_1 + M_2 c_2 = \rho$, eqn (4.17) can be inversed as

$$J_{x,1}^m = (M_1 M_2/V_2 \rho)J_{x,1}^v = -D_{12}(M_1 M_2/V_2 \rho)\frac{dc_1}{dx} \tag{4.18}$$

The last equality is obtained by using eqn (4.7), after recalling that, if fluxes in the volume-fixed reference frame are used, the binary diffusion coefficient, D_{12}, can be used. The mass fraction of component 1, w_1, is given by

$$w_1 = M_1 c_1 / (M_1 c_1 + M_2 c_2) \tag{4.19}$$

Differentiating this equation, with the use of eqn (4.16) gives

$$dc_1 = (V_2 \rho^2 / M_1 M_2) dw_1 \tag{4.20}$$

which, after substitution into eqn (4.18) gives the following diffusion equation for component 1 and, by symmetry, an equally valid equation for component 2:

$$J_{x,1}^m = -D_{12} \rho \frac{dw_1}{dx} \tag{4.21}$$

$$J_{x,2}^m = -D_{12} \rho \frac{dw_2}{dx} \tag{4.22}$$

These equations, including the density ρ, are thus of a slightly different form from eqns (4.7) and (4.8). The advantage of using them is that they both contain the same binary diffusion coefficient as was used previously for the volume-fixed frame of reference, in spite of now referring to the mass-fixed frame.

For multicomponent mixtures, the situation is more complex and the reader is referred to other texts (e.g. Tyrrell and Watkiss 1979). For M components there are $M(M - 1)/2$ independent diffusion coefficients. For dilute solutions of the $M - 1$ remaining components in solution in a fluid substance, which is denoted as component 1, only the $M - 1$ diffusion coefficients D_{1i}, where i has any value between 2 and M, are important.

4.2.2 The diffusion equation

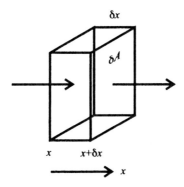

Another equation is now developed for diffusion, which is sometimes known as Fick's second law. The mass-fixed reference frame is used, and again diffusion and concentration gradients in the x-direction only are taken into account. Now two elemental areas of equal area, δA, perpendicular to the x-direction are considered, one at x and one at $x + \delta x$. An elemental volume of $\delta A \delta x$ is formed between the two areas. As the mass-fixed reference frame is being used, the volume may change during diffusion, although the mass it contains will stay constant at its initial value of $\rho \delta A \delta x$. If the mass-fraction gradient of component 1 at x is dw_1/dx, then at $x + \delta x$ it will have changed to

$$\left(\frac{dw_1}{dx}\right) + \left(\frac{d^2 w_1}{dx^2}\right)\delta x \tag{4.23}$$

as the mass-fraction gradient will not be linear in general. The mass of component 1 in the elemental volume $\delta A \delta x$ will be changing at a rate, $\delta m_1/\delta t$, given by the difference in the fluxes at the two faces multiplied by their area δA, which can be calculated from eqn (4.21), to be

$$\frac{\delta m_1}{\delta t} = -D_{12}\rho\left(\frac{dw_1}{dx}\right)\delta A + D_{12}\rho\left[\left(\frac{dw_1}{dx}\right) + \left(\frac{d^2 w_1}{dx^2}\right)\delta x\right]\delta A$$

$$= D_{12}\rho\left[\frac{d^2 w_1}{dx^2}\right]\delta A \delta x \tag{4.24}$$

Equation (4.8) assumes that D_{12} is not changing with concentration. As $m_1 = w_1 \rho \delta A \, \delta x$ and as the δ quantities go to zero, eqn (4.8) can be rewritten as

$$\frac{\partial w_1}{\delta t} = D_{12}\frac{\partial^2 w_1}{\partial x^2} \tag{4.25}$$

with a similar equation for component 2, which is Fick's second law of diffusion. Generalising to three dimensions the equation for both components is

$$\frac{\partial w_i}{\delta t} = D_{12}\nabla^2 w_i \tag{4.26}$$

and if the diffusion coefficient is dependent on position (typically because of dependence on concentration) the equation has to be written as

$$\rho\frac{\partial w_i}{\delta t} = \text{div } \rho D_{12} \text{ grad } w_i \tag{4.27}$$

This equation is valid for a mass-fixed frame of reference and it desirable to have a version which is valid in a laboratory-fixed frame. The problem in one dimension is considered first. Let v_x be the velocity of centre of mass at any point in the fluid at any time. In laboratory-fixed coordinates, the quantity $\partial w_i/\partial t$ will change as the centre of mass moves, also because the concentration is changing with distance. In fact it will decrease by an amount $v_x \partial w_i/\partial x$ and thus

$$\frac{\partial w_i}{\delta t} = D_{12} \frac{\partial^2 w_i}{\partial x^2} - v_x \frac{\partial w_i}{\partial x} \tag{4.28}$$

In three dimensions taking into account the possible variation of D_{12} with concentration, this becomes,

$$\rho \frac{\partial w_i}{\delta t} = \text{div } \rho D_{12} \text{ grad } w_i - \rho \boldsymbol{v} \cdot \text{grad } w_i \tag{4.29}$$

When the concentration of component 2 is very small, so that ρ, $\boldsymbol{v}$, and D_{12} are independent of w_2, $c_2 = \rho w_2/M_2$, and so a useful equation in terms of concentration can be obtained

$$\frac{\partial c_2}{\delta t} = \text{div } D_{12} \text{ grad } c_2 - \boldsymbol{v} \cdot \text{grad } c_2 \tag{4.30}$$

which can be used for some problems, e.g. for chromatography in Section 7.1.1.

4.2.3 The behaviour of the diffusion coefficient at higher density

At the end of Section 4.1, the use of the rigorous kinetic theory of Chapman and Enskog was said to be an accurate way of predicting transport properties at low density and that the Thorne–Enskog extension of this theory gave some success as the density increased (Maitland *et al.* 1981). For higher density, prediction is more difficult, but the approach of non-equilibrium thermodynamics (De Groot and Mazur 1962, Tyrrell and Watkiss 1979) can give some insight into the behaviour of the diffusion coefficient at the higher densities of supercritical fluids. For a fuller understanding of this approach, the reader is referred to specialized textbooks and here the results are used without justification. One of the basic tenets of non-equilibrium thermodynamics is the *fluctuation-dissipation theorem*, which says that the rates at which the gradients of quantities dissipate as a result of the transport properties is related to the rates at which fluctuations in these quantities dissipate in uniform media. In Section 2.5.4 the existence of fluctuations in concentration were introduced. A fluctuation in concentration will disappear in time, on average at a rate related to the diffusion coefficient, according to the theorem. Using this approach, the transport coefficients can be related to phenomenological coefficients, which can be discussed in terms of the behaviour of systems which are uniform on average.

For simplicity, a binary mixture is again considered, with component 1 considered as the solvent substance and component 2 as the solute. The motion of the solute species, component 2, is now considered, being more relevant to later discussions, although equally valid expressions for component 1 can be obtained. Non-equilibrium thermodynamics gives the following expression for the binary diffusion coefficient:

$$D_{12} = \frac{RT}{M_2} \cdot \frac{(x_1 M_1 + x_2 M_2)^2}{x_1 M_1^2} \cdot B_2 t_{c,2} \qquad (4.31)$$

where B_2 is the derivative of the natural logarithm of the fugacity, $\ln(f_2/p^{\ominus})$, with respect to that of the mole fraction, $\ln x_2$, at constant pressure and temperature, i.e.

$$B_2 = \left(\frac{\partial \ln(f_2/p^{\ominus})}{\partial \ln x_2} \right)_{p,T} \qquad (4.32)$$

and $t_{c,2}$ is a correlation time for molecular velocities of the component 2, defined by

$$t_{c,2} = \frac{M_2}{3 N_2 RT} \cdot \int_0^\infty \left\langle \sum_i \sum_j c_{2,i}(0) \cdot c_{2,j}(t) \right\rangle dt \qquad (4.33)$$

The integrand in eqn (4.33) is a so-called *time-correlation function*. It is the vector product of the velocity of a particular molecule of component 2 at an arbitrary time zero and the velocity of all molecules at a later time, summed over all the molecules. It is therefore a measure of the persistence of the velocity of a molecule with time, either by the continuance of its own motion in the general original direction, or by passing on this motion to another molecule. In practice, the transfer of velocity between solute molecules will not usually be important, and the main component of the time-correlation function will be the persistence of the motion of individual molecules. N_2 is the number of molecules in the system and so the value of the correlation function at zero time will be given by the total mean squared velocity of molecules of component 2, $N_2 \langle c_2^2 \rangle$, which is known to be $N_2 3RT/M_2$. As the inverse of this quantity is included as a factor in eqn (4.33), $t_{c,2}$ is equal to the integral of a function which is unity at zero time and falls, as velocity correlation is lost, towards zero. Thus $t_{c,2}$ is a characteristic time for the fall of the velocity correlation.

4.2.4   Dependence of the diffusion coefficient in dilute solution on density

The behaviour of diffusion coefficients is first considered for a very dilute binary solution of component 2 in the fluid substance component 1, for which the quantity $(x_1 M_1 + x_2 M_2)/x_1 M_2^2$ can be considered to be unity. As will be shown in the next section when B_2 will be considered more fully, for dilute solutions also $B_2 \to 1$. Equation (4.31) then simplifies to

$$D_{12} = \frac{RT t_{c,2}}{M_2} \tag{4.34}$$

At constant temperature, the correlation time is expected to depend inversely on density, as the distance a molecule can travel before collision, and therefore the time on average before its velocity is disrupted, would be expected to have an inverse density relationship. The open circles in Fig. 4.1 show some experimental measurements (Feist and Schneider 1982) for the diffusion coefficient of naphthalene in carbon dioxide at 313 K, multiplied by density and then plotted against density. These show an approximately inverse relationship between diffusion coefficient and density and other published measurements also fall near the same curve. The solid point near zero density is the value calculated for 1 bar using rigorous kinetic theory, using an experimental measurement of the diffusion coefficient for naphthalene in air to obtain σ for naphthalene (Mack 1925). The dotted line emanating from this point was calculated from Thorne–Enskog theory. It can be seen that the product ρD_{12} for both the experimental and calculated values stays within

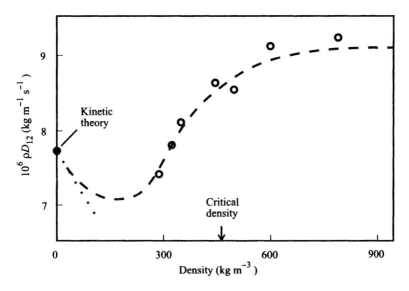

Fig. 4.1   The product of the density and diffusion coefficient for naphthalene at 313 K, plotted versus density.

a 25 per cent band over the density range and the dashed curve shows the general trend.

A possible explanation of the rising part of the curve can be made using the concept of 'edgewise' diffusion. This concept is used to explain the fact that flat molecules, such as benzene diffuse more rapidly in viscous fluids than would be predicted by hydrodynamic theories based on their behaviour in more mobile fluids (Tyrrell and Watkiss 1979). It is argued that when rotational diffusion of molecules becomes slow, translational movements persist more strongly along particular molecular axes. For a non-spherical molecule, molecular movement will have less resistance in certain directions. The vector product in eqn (4.33) is therefore enhanced because the velocity components along favoured axes will correlate for a longer period of time, thus increasing the diffusion coefficient. The onset of edgwise diffusion as the density is increased, may thus account for the rising part of the dashed line in Fig. 4.1. Whether this explanation is correct or not, the discussion illustrates the difficulty in predicting diffusion coefficients in dense media.

4.2.5   Critical region effects on the diffusion coefficient in less dilute solutions

When the approximations made in the last section for very dilute solutions are not valid, eqn (4.31) rather than eqn (4.34) must be used. It is found that as concentration is increased, the quantity B_2 is the first of the factors in the equation to begin to depart from unity significantly, particularly near the critical temperature and density. Here some qualitative calculations, based on the van der Waals equation, are presented to illustrate the general behaviour. Equation (2.15) gives the van der Waals expression for the fugacity in a binary mixture and, after differentiation, the following equation is obtained for B_2

$$B_2 = 1 + \left(\frac{\partial V}{\partial P}\right)_{T,x} \frac{2x_1 x_2 B_2'}{V^2 (V-b)^2} \tag{4.35}$$

where $(\partial V/\partial p)_{T,x}$ is that given by eqn (2.8) and

$$B_2' = (a_{11} + a_{22} - 2a_{12})V + \frac{a(b_1 - b_2)^2}{V}$$
$$- 2[(a_{11} - a_{12})x_1 - (a_{22} - a_{12})x_2](b_1 - b_2) \tag{4.36}$$

Plots of this function versus pressure at three temperatures, 1 K, 3 K, and 9 K above the critical temperature, for a solution of 4×10^{-4} mole fraction of naphthalene in carbon dioxide, are shown in Fig. 4.2. The calculations were made using the van der Waals parameters given in Section 2.5.3. As can be seen the function departs from unity near the critical density to an

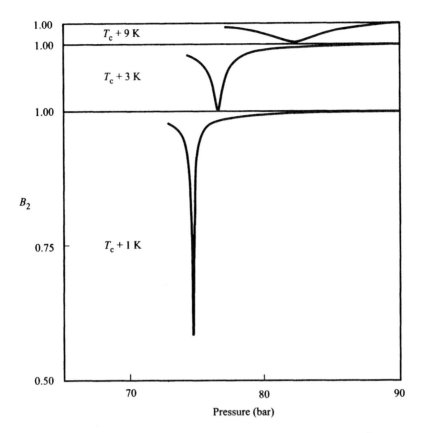

Fig. 4.2 The quantity $B_2 = (\partial \ln f_2/p^{\ominus}/\partial \ln x_2)_{p,T}$ for naphthalene ($x_2 = 4 \times 10^{-4}$) in carbon dioxide at various temperatures above the critical temperature of carbon dioxide, plotted versus pressure.

increasing extent as the critical temperature is approached. As the concentration of naphthalene is reduced, the departures from unity decrease approximately in proportion to the concentration, so that as $x_2 \to 0$, $B_2 \to 1$, as was asserted in the last section.

Thus for quite a modest concentration, the diffusion coefficient is reduced by almost a factor of two due to this effect. As the critical point is approached closely, the diffusion coefficient tends to zero for finite concentrations. A physical explanation of this is that, as the temperature is lowered towards the critical temperature at the critical density, a situation is being approached where two phases with two different concentrations of the solute coexist in equilibrium, and where there is no tendency to reduce the concentration difference. Some experimental observations have been made of this effect (e.g. Saad and Gulari 1984). Conversely, at constant higher density, there is undramatic behaviour of the diffusion coefficient as the temperature is lowered from supercritical conditions, through the critical temperature, to liquid conditions (e.g. Lauer *et al.* 1983).

4.2.6   The effect of moving frames of reference in the critical region

The basic equation for treating a diffusional problem in a supercritical fluid is eqn (4.29), which can be very difficult to solve rigorously, as ρ, v, and D_{12} are all dependent on the mass fractions, i.e. w_2. The relationship of density to mass fraction,

$$\rho = 1/[w_2(M_2 V_2 - M_1 V_1) + M_1 V_1] \tag{4.37}$$

is straightforward and the dependence of D_{12} on concentration, albeit in terms of mole fractions, has been discussed in the last section. In this section, the dependence of v on concentration will be discussed in general terms, at constant temperature and pressure. The discussion starts from the continuity equation for the fluid,

$$\rho \operatorname{div} v = -\partial \rho/\partial t - v \cdot \operatorname{grad} \rho \tag{4.38}$$

which, as ρ only depends on w_2 at constant pressure and temperature, can be rewritten, after dividing by ρ, as

$$\operatorname{div} v = -(\partial \ln \rho/\partial w_2)[(\partial w_2/\partial t) + v \cdot \operatorname{grad} w_2] \tag{4.39}$$

which, using eqns (4.29) and (4.37), becomes

$$\operatorname{div} v = \rho(M_2 V_2 - M_1 V_1)\operatorname{div} \rho D_{12} \operatorname{grad} w_2 \tag{4.40}$$

This equation can thus be used to calculate the velocity of the barycentric reference frame, with respect to which diffusion takes place according to the equations of Section 4.2.1, and laboratory-fixed coordinates. When diffusion takes place in one direction, eqn (4.40) can be integrated with a suitable boundary condition to give an expression for v. This is carried out to analyse accurate experiments of diffusion coefficients in liquids (Tyrrell and Watkiss 1979).

In supercritical fluids effects of reference-frame movements can be very important in the critical region, because partial volumes can be large, as described in Section 2.5.3. A graphic illustration of this effect is a report of an experiment (Weiser 1957) in which iodine was dissolved in supercritical carbon dioxide in a view cell. After dissolving, the iodine diffused into the bulk of the solution, but near the critical density this diffusion appeared to stop completely. The experiment is illustrated schematically in Fig. 4.3. This shows a vessel containing supercritical carbon dioxide near the critical density and

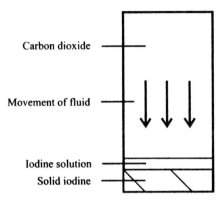

Fig. 4.3 Schematic illustration of the solution and diffusion of iodine in supercritical carbon dioxide.

Table 4.3 The diffusion coefficient, D_{12}, of benzene in dilute solution in water at 250 bar at subcritical and supercritical temperatures (Carroll 1991)	Temperature (°C)	Density (kg m⁻³)	$10^9 D_{12}$ (m² s⁻¹)
	300	743	7.4
	350	626	25
	380	459	66
	400	166	160

temperature, in which some solid iodine lies on the bottom. The iodine dissolves to form a thin layer of solution above the solid. The partial molar volume of iodine under these conditions is large and negative and so the density of the iodine solution is much larger than that of the bulk carbon dioxide. To ensure that this is the case, as the iodine dissolves, carbon dioxide must move downwards through the vessel, as shown. Consequently the upper visible boundary of the iodine tends to move downward, counteracting the upward movement by diffusion. This boundary therefore appears to be stationary. The reluctance of the iodine to diffuse upwards in the critical region will be partly because the diffusion coefficient is lower, as described in the last section, but mainly because of the movement of the barycentric frame of reference.

4.2.7 Diffusion in supercritical water

Measurements of the diffusion coefficient of benzene in supercritical water by chromatographic peak broadening show that the diffusion increases rapidly as the temperature is raised. Results at a constant pressure are given in Table 4.3 (Carroll 1991). This rise is associated with a fall in water density, but it is more dramatic, and it is likely that part of the increase is due to the breakdown of water structure in the supercritical region. as discussed in Section 1.4.1.

4.3 Viscosity

At low pressures, below one atmosphere, the (dynamic) viscosity, η, of a gas is approximately constant, but thereafter rises with pressure in a similar way to density, ρ. However, the dependencies of density and viscosity on pressure at constant temperature are not conformal. Of interest therefore is the kinematic viscosity, η/ρ, which is illustrated in Fig. 4.4 for carbon dioxide (Bartle *et al.* 1989). At constant temperature, kinematic viscosity falls from high values at low pressure until the critical density and then rises slightly. As well as illustrating the comparative behaviour of dynamic viscosity and density, the kinematic viscosity is proportional to the pressure drop through a non-turbulent system for a given mass flow rate. For a uniform tube of radius a, with gas flowing through at a given mass flow rate of m, the pressure variation with length, l, along the tube is given by

$$\mathrm{d}p/\mathrm{d}l = -(8m/\pi a^4)(\eta/\rho) \qquad (4.41)$$

A comprehensive correlation for the viscosity of carbon dioxide has been published (Vesovic *et al.* 1990).

The viscosity of supercritical water exhibits somewhat different behaviour, as shown in Fig. 4.5, where η/ρ is plotted this time against density (Haar *et al.* 1984, Watson *et al.* 1980). The graph shows that at higher densities the

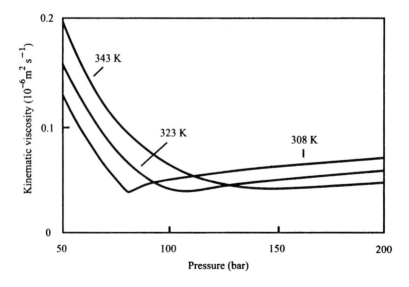

Fig. 4.4 Isotherms of the kinematic viscosity of carbon dioxide as a function of pressure.

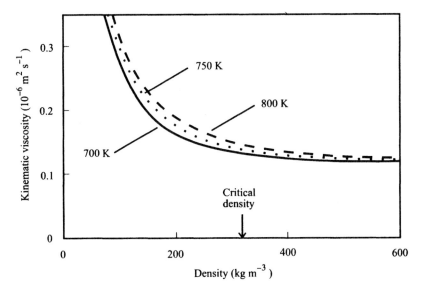

Fig. 4.5 Isotherms of the kinematic viscosity of water as a function of density.

kinematic viscosity tends towards a value of approximately $0.12 \times 10^{-6} \, \mathrm{m^2 \, s^{-1}}$ at all temperatures in the range. Comparison of Figs 4.4 and 4.5 shows that the kinematic viscosities of carbon dioxide and water are not widely different; another indication of the loss of water structure under supercritical conditions.

4.4 Thermal conductivity

Like viscosity, thermal conductivity rises with density. Isotherms of the thermal conductivity of carbon dioxide are shown in Fig. 4.6 (Johns *et al.* 1986). The scale is the same over the whole *y*-axis, but it has been broken so that the isotherms at 302 K, 316 K, 348 K, 380 K, and 430 K are plotted 60, 50, 30, 20, and 10 units below the isotherm at 470 K, to avoid confusing the data. As well as the monotonic rise with density, thermal conductivity also shows enhancement in the critical region, and this is quite marked even 12 K above the critical temperature, as shown on the isotherm for 316 K in Fig. 4.6. This enhancement is related to that in the heat capacity at constant pressure, as discussed in Section 1.4.

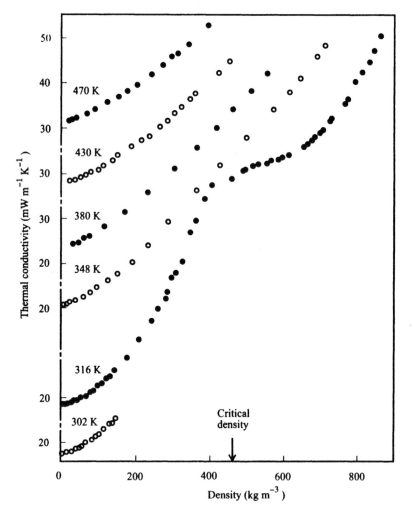

Fig. 4.6 Isotherms of the thermal conductivity of carbon dioxide as a function of density.

5 *Extraction*

5.1 Introduction

Supercritical fluid extraction (SFE) is becoming an important process, as well as a tool in analytical science, and has seen rapid development in the last few years. A number of plants are now operating. It has the advantages, compared with liquid extraction, that

(1) it is usually more rapid;

(2) the solvent is easier to remove;

(3) pressure (as well as temperature and the nature of the solvent) can be used to select, to some extent, the compounds to be extracted; and

(4) carbon dioxide is available, to be used as a pure or modified solvent, with its convenient critical temperature, its cheapness and non-toxicity.

This chapter explains the principles of SFE on a small scale in terms of the physical chemistry of the process, which is different to, but has analogies to the methods used by chemical engineers to model SFE on a process scale. For the most part the extraction of a single compound is considered, and if more than one compound is being extracted, the behaviour of each of them is assumed to be independent. However, mixtures and their mutually dependent effects are considered in Section 5.5.

The basic process of extraction is shown schematically in Fig. 5.1, which represents both large process scale and small laboratory scale for process development or analysis. The fluid substance, such as carbon dioxide, is pumped as a liquid, and therefore is initially cooled to, say, 5 °C, which must allow for some heating during pumping. This cooling will take place in a reservoir, as shown, for pilot or process scale, but on a small scale it may be done by cooling the pump head. A system for adding a proportion of liquid modifier, not shown, may be incorporated. The fluid is then heated to the extraction temperature and pumped into an extraction cell, which is maintained at this temperature. On a small scale, a coil of tubing followed by an extraction cell may be incorporated into a controlled oven. The matrix to be extracted will be

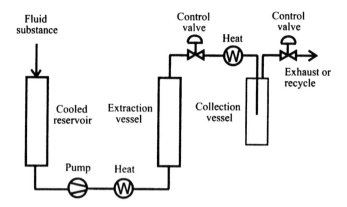

Fig. 5.1 Schematic diagram of a system for supercritical fluid extraction.

packed into the extraction cell in a mesh basket or between frits to prevent it being carried out of the cell during extraction.

Following extraction, the pressure is reduced to precipitate the extract through a control valve, or sometimes on a small scale through a section of capillary tubing. The flow rate of fluid is controlled by the rate of pumping and the pressure in the extraction cell is controlled by the setting of the control valve for a particular pumping rate. Control systems may be used to control the extraction conditions. Reduction of pressure causes cooling of the fluid and so heat input is required, as shown. The precipitated material is collected at the base of the collection vessel, which has temperature control and also pressure control from the control valve on the fluid exit. A series of collection vessels at successively lower pressures may be employed as discussed below and shown below in Fig. 5.19.

For small scale extraction, the collection pressure is likely to be atmospheric pressure and a collection solvent may be placed in the collection vessel to more effectively trap the product, ready for analysis. A trapping liquid, such as a vegetable oil, may also be used on a process scale to give a particular product. Trapping on to a surface, such as that of active charcoal, may also be used, particularly for volatile products, followed by thermal desorption. On a process scale the fluid leaving the collection vessel will be cooled for recycling, but on a smaller scale will be discarded. The thermodynamic processes occurring during a supercritical fluid extraction process are illustrated in Fig. 1.8.

The process of extraction can be considered to involve the three factors shown in the SFE triangle below. The solute must, firstly, be sufficiently soluble in the supercritical fluid to be removed by solution in the fluid flow. If this is not the case, it will be revealed by interpretation of the kinetic recovery

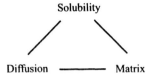

curve, as will be shown below. Secondly, the solute must be transported sufficiently rapidly, by diffusion or otherwise, from the interior of the matrix in which it is contained. The latter 'diffusion' process may be normal diffusion of the solute, as in a polymer, or it may involve diffusion in the fluid through pores in the matrix. The time-scale for diffusion will depend on the diffusion coefficient and the shape and dimensions of the matrix or matrix particles. Of these the shortest dimension is of great importance, as the times depend on the square of its value. Thirdly, the solute must be released by the matrix. This last process may involve desorption from a matrix site, passage through a cell wall, or escape from a cage formed by polymer chains. It can be slow and in some cases it appears that part of the substance being extracted is locked into the structure of the matrix. A related problem is the presence of water. Water is not very soluble in many fluids, such as carbon dioxide, and it can 'mask' the substances to be recovered. The rate of extraction may sometimes be equal to the rate of water removal. It may be necessary to dry the material to be extracted in air or by admixture with a drying agent. Reduction of the water content of plant material from, say, 80 per cent (as measured by mass loss at 100 °C) towards 10 per cent may be desirable, provided valuable volatiles are not lost in the process. However, water may assist extraction by acting as a modifier, as is believed to be the case for coffee decaffeination.

Modifiers or entrainers added to the fluid, as discussed in Section 2.1.1, may be beneficial to any of the above factors. They may improve the solubility of the compounds to be extracted, and this was originally thought to be their most important rôle. However, they often improve diffusion by absorption into a polymer, for example. Modifiers may improve the matrix factor by adsorbing on surface sites. Modifiers, such as methanol, can reduce the water problem by improving its solubility in the fluid.

Figure 5.2 shows examples of the types of curves of recovery versus time that can be obtained in SFE. (a) is a typical curve obtained when the process is controlled by diffusion. When matrix effects are significant, the results may have the form of (b). Curve (c) is an example of recovery behaviour when the extracted compound is not very soluble in the extracted fluid.

5.2 Modelling of SFE

Four steps are considered in the models described in this chapter:

(1) rapid *fluid entry* into the matrix;

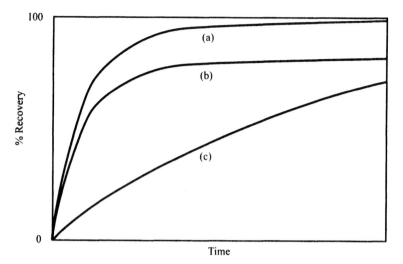

Fig. 5.2 Examples of recovery curves: (a) is a typical diffusion-controlled curve; (b) one showing significant matrix effects; and (c) one for a poorly soluble solute.

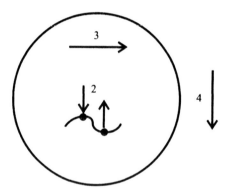

Fig. 5.3 Steps in the SFE process: (1) rapid *fluid entry* into the matrix (not shown); (2) a *reversible release* process such as desorption from matrix sites or penetration of a biological membrane; (3) *transport*, by diffusion or otherwise to the edge of a matrix particle; and (4) *removal in solution* in the fluid.

(2) a *reversible release* process such as desorption from matrix sites or penetration of a biological membrane;

(3) *transport*, by diffusion or otherwise to the edge of a matrix particle; and

(4) *removal by solution* in the fluid.

Figure 5.3 illustrates steps (2)–(4) in the process. Step (1) is considered to be too fast to significantly affect the kinetics of recovery. For most of this chapter steps (3) and (4) are considered, as most experimental data have been analysed by discussing these effects. The effect of the matrix has not been much discussed quantitatively, and so only near the end of the chapter, in Section 5.6, are steps (2), (3) and (4) all taken into consideration. The equations for the involvement of both steps (3) and (4) are developed and discussed in Section 5.3. The two limits of these equations when steps (3) or (4) become predominant are considered in Sections 5.4 and 5.5, respectively.

5.3 Control by transport through the matrix and removal by solution

5.3.1 Modelling for a sphere with uniform initial concentration

A collection of identical spherical particles of radius a are considered and are treated as spherically symmetrical, with the distance from their centres given by r. The concentration of the compound to be extracted within the sphere, c, will be a function of both r and time, t, i.e. $c \equiv c(r, t)$. The diffusion equation within the sphere is then solved, which in spherical coordinates is

$$\partial cr/\partial t = D(\partial^2 cr/\partial r^2) \tag{5.1}$$

where D is the diffusion coefficient or effective diffusion coefficient of the compound within the sphere, which is assumed to be constant over the sphere. The spheres are considered to be in a fluid, which is passing through them with a volume flow rate of F. It is assumed that the concentration of solute in the fluid is constant throughout at c_f, because of rapid diffusion and convection within the fluid. The rate diffusion of the compound out of the spheres is then equated to the rate of its removal by solvation in the fluid, i.e.

$$-AD(\partial c/\partial r)_{r=a} = Fc_f \tag{5.2}$$

where A is the total surface area of all the spheres.

It is now assumed that the concentration in the fluid is proportional to the concentration in the matrix at the surface at a, the proportionality constant being the partition coefficient, K, defined as a ratio of concentrations or amounts per unit volume and thus

$$c_f = Kc(a, t) \tag{5.3}$$

Combining eqns (5.2) and (5.3) gives

$$-(\partial c/\partial r)_{r=a} = hc(a, t) \tag{5.4}$$

where

$$h = KF/AD \tag{5.5}$$

If K or F are large and D is small, removal by solution in the solvent will not be rate determining. Conversely, if D is large and K or F are small, transport out of the matrix will not be rate determining. Thus the larger the value of ha, the more important transport is in determining the rate of extraction, whilst for smaller values of ha solvation in the fluid and removal by the fluid flow becomes more rate determining.

Initially it is assumed that concentration is constant over the volume of the sphere, i.e. that

$$c(r, 0) = c_0 \tag{5.6}$$

Equation (5.1) is now solved in conjunction with the boundary conditions to obtain an expression for the concentration as a function of position within the sphere and time, $c(r, t)$. Solutions of the equation have been previously published in connection with heat conduction (Carslaw and Jaeger 1959) and related diffusion (Crank 1975) problems. The expression for $c(r, t)$ will not be given here, but instead that for the total mass of solute within the sphere at a given time

$$m = \int_0^a 4\pi r^2 c\,(r, t)\mathrm{d}r \tag{5.7}$$

divided by the initial mass in the sphere, $m_0 = (4/3)\pi a^3 c_0$. This ratio is given by

$$m/m_0 = 6 \sum_{n=1}^{\infty} (ha/\lambda_n)^2 [ha(ha - 1) + \lambda_n^2]^{-1} \exp[-(\lambda_n/\pi)^2 t/t_{\mathrm{c}}] \tag{5.8}$$

where

$$t_{\mathrm{c}} = a^2/\pi^2 D \tag{5.9}$$

and λ_n are the roots of the equation

$$\lambda \cot(\lambda) = 1 - ha \tag{5.10}$$

These roots are tabulated or are available from computer library routines and, in the limit of large values of ha are given by

$$\lambda_n = n\pi \tag{5.11}$$

Plots of values of $\ln(m/m_0)$, obtained from eqn (5.8), are shown in Fig. 5.4 for various values of ha. In each case, the curve falls to a linear portion at longer times. This is because the higher terms in the sum of exponentials in eqn (5.8) become less important as time increases and $\ln(m/m_0)$ is given approximately by

$$\ln(m/m_0) = \ln\{6(ha/\lambda_1)^2 [ha(ha - 1) + \lambda_1^2]^{-1}\} - (\lambda_1/\pi)^2 t/t_{\mathrm{c}} \tag{5.12}$$

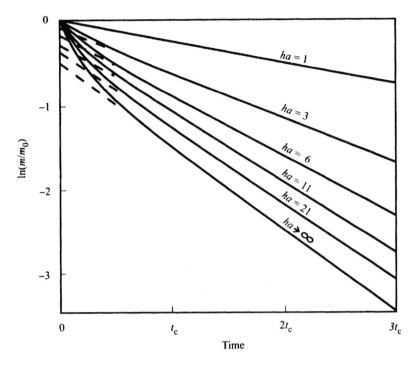

Fig. 5.4 Theoretical curves for SFE of a sphere, including the effects of transport and removal in solution, for different values of the parameter *ha.*

The slopes of the linear portions of the graphs are thus equal to $-(\lambda_1/\pi)^2/t_c$ and the extrapolated intercepts, shown by the dashed lines in Fig. 5.4, $-I$, are given by

$$-I = \ln\{6(ha/\lambda_1)^2[ha(ha-1) + \lambda_1^2]^{-1}\} \qquad (5.13)$$

Values of λ_1 and I are given in Table 5.1 for the values of *ha* which were used in the plots of Fig. 5.4. It can be seen that both these quantities fall as *ha* falls, reducing both the high rate at the beginning of the extraction and the slower rate corresponding to the slope of the linear portion. The curves are plotted against time in units of t_c and so they compare the effects of partitioning and flow rate for a constant diffusion coefficient. The curve for *ha* tending towards infinity corresponds to the situation where only transport is important and this is discussed below in Section 5.4.1. As *ha* falls removal by solution in the fluid becomes more and more rate determining.

A physical interpretation of the process is illustrated in Fig. 5.5, which shows how the concentration profile changes during extraction. Initially in (a), it is

Table 5.1 Parameters for the spherical model, including the effects of transport and removal in solution

ha	λ_1	I	Recovery (per cent) after $0.3 t_c$
∞	3.1416	0.4997	63
21	2.9930	0.3731	57
11	2.8628	0.2884	52
6	2.6537	0.1887	46
3	2.2889	0.0866	36
1	1.5708	0.0146	23

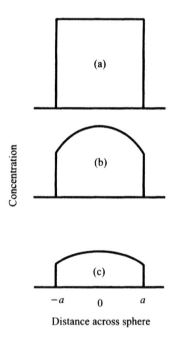

Fig. 5.5 Concentration profiles across a sphere during SFE involving the effects of transport and removal in solution.

constant across the sphere. Passage to the profile shown in (b) corresponds to the non-linear portion of the curves in Fig. 5.4. Once this profile is established, it reduces in size but maintains the same shape, as shown in (c), during the final exponential decay. If ha is large, the vertical portion of the profiles in (b) and (c) are very small and the non-exponential part of the extraction curve is more important. If ha is small, the curved portion of the profiles in (b) and (c) are very flat and nearly the whole extraction curve is exponential.

Plots of percentage recovery versus time, drawn from the same equations, are shown in Fig. 5.6 for various values of *ha*. For *ha* = 1, representing limitation by partitioning into the fluid, a slow recovery of exponential form is obtained. As *ha* is increased, the rate of recovery rises and the form changes to that of diffusion control. However, raising *ha*, by increasing solubility or flow rate, has diminishing returns, as when diffusion control takes over, increasing *ha* has little effect. Thus the curves for *ha* = 30 and *ha* = 100 are very similar. The curves are plotted versus time in terms of t_c and the relationship to real time is given by the parameter D/a^2 using eqn (5.9). Thus, if experimental data are fitted to the theoretical curves, either recovery curves or logarithmic plots, the two parameters *ha* and D/a^2 are obtained. The total volume, *V*, of the particles is related to the total surface area by

$$V = (a/3)A \tag{5.14}$$

Therefore the product of *ha* and D/a^2 is given by

$$(ha)(D/a^2) = KF/3V \tag{5.15}$$

which allows a value of *K* to be obtained, provided the volume of the matrix and the flow rate are known.

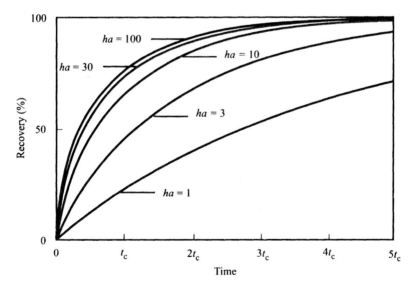

Fig. 5.6   Plots of the percentage recovery during SFE of a sphere, as a function of time for different values of *ha*.

5.3.2 Comparison of the spherical model with experimental results

If the flow rate is varied at constant pressure and temperature for SFE from a polymer, D/a^2 is expected to remain constant, whereas ha is expected to rise in proportion to the volume flow rate, F. Data for the SFE of the combined amounts of m- and p-xylene form polystyrene beads, varying in size from 0.18 mm to 2.0 mm diameter, for various flow rates (Hawthorne *et al.* 1995) were fitted to the appropriate equations and the comparison shown in Fig. 5.7. (The flow rates were measured as liquid CO_2 at the pump, but will be proportional to the fluid flow rate in the extraction cell.) For all the theoretical curves values of $D/a^2 = 0.0009$ and $ha = 16(F/\mathrm{ml\ min}^{-1})$ were used. Thus only two parameters were used to fit the curves, and there is qualitative agreement, bearing in mind that the sample did not consist of spheres of uniform size as strictly required by the theory.

If the pressure is varied at constant flow rate and temperature, both the parameters D/a^2 and ha are expected to change. Thus the recovery curves must be fitted for individual pressures, and this has been done for the extraction of Irgafos 168 (tris-(2,4-di-*tert*-butyl) phosphite) from polypropylene at various pressures, as shown in Fig. 5.8. The particles were irregular spheres of diameter 0.8 ± 0.2 mm and extraction was carried out at 45 °C with pure CO_2 at a flow rate of 7 ml s^{-1}, measured with a bubble flow meter at 20 °C and 1 bar (Bartle *et al.* 1991*b*). Fitting is now much better and the parameters obtained from the

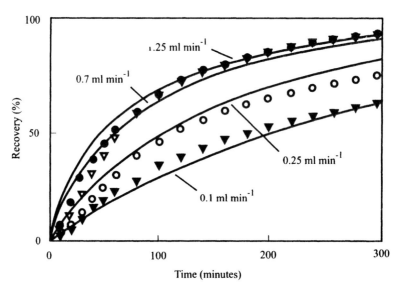

Fig. 5.7 Comparisons of experimental data and model predictions (continuous lines) for SFE of *m*- and *p*-xylene from polystyrene beads at various flow rates.

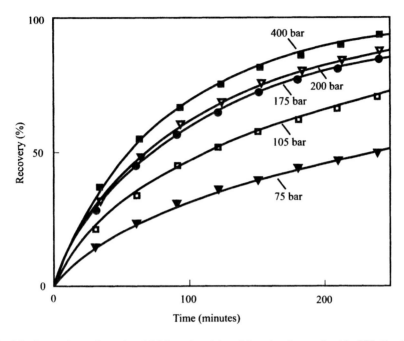

Fig. 5.8 Comparisons of experimental data and model predictions (continuous lines) for SFE of Irgafos 168 (tris-(2,4-di-*tert*-butyl) phosphite) from polypropylene at various pressures.

Table 5.2 Values of the parameters obtained by fitting the data shown in Fig. 5.8	Pressure (bar)	$D/a^2 \times 10^5$ (s^{-1})	ha
	75	21	3.2
	105	48	5.8
	175	90	7.3
	200	100	8.1
	400	160	8.2

fitting are given in Table 5.2. The values of ha are also shown in Fig. 5.9, plotted against pressure, and can be seen to have the same form as a solubility curve versus pressure (Fig. 3.1). This is to be expected as K is proportional to solubility and, therefore, so will h be. The values of D/a^2 in Table 5.2 also rise with pressure and this is explained by higher absorption of the supercritical fluid substance at higher pressures causing the polymer to swell, raising the diffusion coefficient. Thus with polymers, increasing the pressure can be beneficial to SFE, even above pressures where the solubility is no longer rising.

The effect of pressure on SFE, due to its influence on solubility, is well known. It is most obvious if extractions are carried out for a particular time.

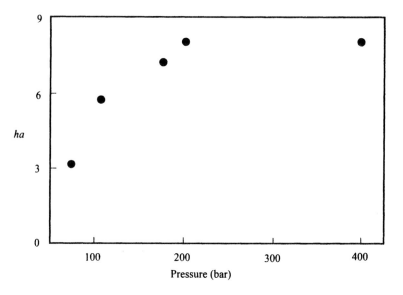

Fig. 5.9   Values of the parameter *ha* obtained by analysis of the data in Fig. 5.8.

Table 5.1 gives the percentage recovery, predicted by the model for a period of $0.3t_c$, for various values of *ha*, which is proportional to solubility. Although the relationship is by no means linear, there is a correlation between *ha*, and therefore solubility, with the amount extracted. Figure 5.10 shows the solubility of atrazine, predicted by the Peng–Robinson equation of state, in (a), and beneath in (b) the experimental percentage recovery of atrazine from soil, both as a function of pressure. The SFE was carried out at 80 °C for 15 minutes using pure CO_2 at a constant flow rate of $5\,\mathrm{ml\,s^{-1}}$ measured with a bubble flow meter at 20 °C and 1 bar (Ashraf *et al.* 1992). This is an example of a so-called pressure threshold curve for SFE.

5.3.3   Modelling for a film
The appropriate model for a film, for example a polymer film, is an infinite slab of thickness L, on the basis that the surface dimensions of the film are far larger than this thickness. Using the corresponding equations that were used in the last section for a sphere, gives the following equation for extraction from an infinite slab with uniform initial concentration distribution:

$$m/m_0 = 8 \sum_{n=0}^{\infty} (hL/\lambda_n)^2 [hL(hL + 2) + \lambda_n^2] \exp[-(\lambda_n/\pi)^2 t/t_c] \qquad (5.16)$$

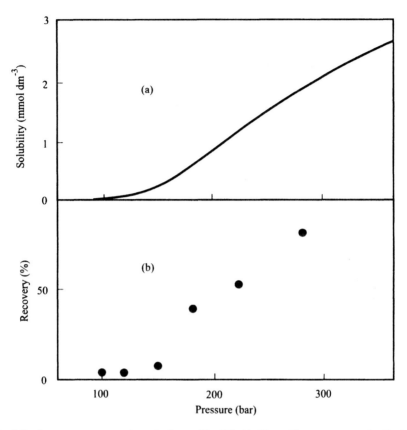

Fig. 5.10   Percentage recovery of atrazine from soil by SFE with CO_2 at different pressures after 15 minutes at 80 °C and constant flow rate, compared with predicted solubility at the same temperature.

where t_c is now given by

$$t_c = L^2/\pi^2 D \tag{5.17}$$

and λ_n are now the roots of the equation

$$\lambda \tan(\lambda) = ha \tag{5.18}$$

Again these roots are tabulated or are available from computer library routines and, in the limit of large values of hL are given by

$$\lambda_n = (2n - 1)\pi \tag{5.19}$$

Equation (5.16) will not be discussed further, except to say that it gives a series of curves of the general form of Fig. 5.4.

5.4   Control by transport through the matrix only

5.4.1   Limiting equations for a sphere

When the compound being extracted is very soluble in the supercritical fluid, it is likely that the partition coefficient, K, will be high and favourable to extraction, and in this case recovery will be controlled by diffusion out of the matrix. The important feature is that h, defined by eqn (5.5) is high and as $ha \to \infty$, $\lambda_n \to n\pi$, as given in eqn (5.11), and eqn (5.8) becomes

$$m/m_0 = \frac{6}{\pi^2} \sum_{n=1}^{\infty} \frac{1}{n^2} \exp(-n^2 t/t_c) \qquad (5.20)$$

where t_c is given by eqn (5.9). Equation (5.20) can be obtained more directly by replacing eqn (5.4) by

$$c(a, t) = 0 \qquad (5.21)$$

indicating that partition into the fluid is so favourable that the concentration of the compound to be extracted is always zero at the surface. Equation (5.1) can then be solved with eqns (5.6) and (5.21) as boundary conditions to give eqn (5.20).

This is perhaps the simplest model for SFE and so it will be discussed in some detail. It can be rewritten as

$$m/m_0 = \frac{6}{\pi^2} \left[\exp(-t/t_c) + \frac{1}{4}\exp(-4t/t_c) + \frac{1}{9}\exp(-9t/t_c) + \cdots \right] \qquad (5.22)$$

The solution is thus a sum of exponential decays, in which at longer times the later (more rapidly decaying) terms will decrease in importance and the first exponential term in the square brackets will become dominant. This can be seen again if the natural logarithm of this equation is taken, after factorizing the term $\exp(-t/t_c)$ from the square bracket, to obtain:

$$\ln(m/m_0) = \ln(6/\pi^2) - t/t_c + \ln\left[1 + \frac{1}{4}\exp(-3t/t_c) + \frac{1}{9}\exp(-8t/t_c) + \cdots \right] \qquad (5.23)$$

The term $\ln(6/\pi^2)$ is equal to -0.4977 and the final term in this equation equals $+0.4977$ at $t = 0$, and so, as required, at $t = 0$, $\ln(m/m_0)$ is also equal to zero. A plot of $\ln(m/m_0)$ versus time therefore tends to become linear at longer times, when the last term in eqn (5.23) tends to zero, and $\ln(m/m_0)$ is given approximately by

$$\ln(m/m_0) = -0.4977 - t/t_c \qquad (5.24)$$

Figure 5.11 is a plot of eqn (5.23), and the straight line portion, which is continued to the $t = 0$ axis as a dashed line, is a plot of eqn (5.24). It is characterized by a relatively rapid fall to a linear portion, corresponding to an exponential 'tail'. The physical explanation of the form of the curve is that the initial portion is extraction, principally out of the outer parts of the sphere, which establishes a smooth concentration profile across each particle, peaking at the centre and falling to zero at the surface. When this has happened, the extraction becomes an exponential decay. The slope of the linear portion is $-1/t_c$ and the linear portion appears to begin at approximately $0.5t_c$. t_c is theoretically related to the effective diffusion coefficient out of the matrix, D, and the radius of the sphere, a, by eqn (5.9). The value of the effective D will usually not be known, although its order of magnitude may be commented on. Most measurements published for D are for true diffusion and for small molecules in relatively mobile solvents (Tyrrell and Watkiss, 1979) and D is of the order of $10^{-9}\,\mathrm{m}^2\,\mathrm{s}^{-1}$. For systems of interest to SFE, D will be of between 1 (for oils) and 4 (for solids) orders of magnitude below this value. For example, values for various solutes in polymers have been given which are of the orders 10^{-11} and 10^{-12}. Equation (5.9) shows a squared dependence on a and rationalizes the common-sense rule that for rapid extraction matrix particles must be small. This may be achieved for solids by crushing or grinding and for liquids by coating on a finely divided substrate or spraying or mechanical agitation. For solid matrix particles with r of the order of 0.1 mm, typical values of t_c are between 10 and 100 minutes.

Figure 5.12 shows some experimental results for the extraction of 1,8-cineole from crushed rosemary at 50 °C using CO_2 at 400 bar (Bartle *et al.* 1990*a*). Many of the compounds extracted from rosemary are highly soluble in

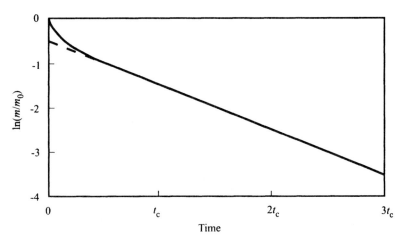

Fig. 5.11 Theoretical curve for the dynamic SFE of a sphere, where extraction is controlled by transport.

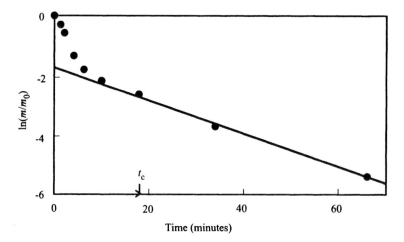

Fig. 5.12 Continuous extraction of 1,8-cineole from crushed, dried rosemary with CO_2 at 50 °C and 400 bar.

carbon dioxide and so the extraction curves are expected to be controlled by transport and in practice they are found to be independent of flow rate. Extraction was continued almost until exhaustion to allow the calculation of values of m and m_0. Similar curves are obtained for the extraction of five other major compounds from rosemary (α-pinene, camphor, camphene, borneol, and bornyl acetate). The experimental results are consistent with the theoretical curve in that the points are close to a straight line after a time of approximately $0.5t_c$. t_c has a value of about 18 minutes in this case, which is obtained from the slope of the straight line portion (it is the time taken for the line to fall one natural logarithm unit). However, the curve differs from the theoretical curve of Fig. 5.11 in that the intercept is less than the theoretical value of -0.5 and this is discussed in the next section.

5.4.2 Limiting equations for a sphere with non-uniform initial distribution

The data shown in the last section exhibited a value of I greater than the theoretical value for a sphere of 0.5. This situation is quite common for the extraction of plant materials, soils, and other matrices. When removal by solvation is partially rate determining, this reduces the value of I as shown in Table 5.1. Another explanation must be sought, and a possibility is that the compound to be extracted is located preferentially near the surface of the matrix particles, which is physically reasonable for these matrices. This concentration distribution will make the initial, more rapid phase of the extraction more important.

Thus, in this section, the compound is considered to be distributed not uniformly, but with concentration falling off exponentially from the surface, i.e.

$$c(r, 0) = c_0 \exp[(r - a)/\alpha] \qquad (5.25)$$

where α is a distance parameter giving the steepness of the fall-off in concentration. A schematic representation of this concentration profile is given in Fig. 5.13. Equation (5.25) now replaces the boundary condition eqn (5.6). Equation (5.1) is now solved with the boundary conditions eqns (5.21) and (5.25), with the solutions being available for adaptation from the sources quoted earlier, to give

$$m/m_0 = \frac{2(a/\alpha)^3}{(a/\alpha)^2 - 2(a/\alpha) + 2[1 - \exp(-a/\alpha)]} \sum_{n=1}^{\infty} \frac{\exp(-n^2 t/t_c)}{n^2 \pi^2 + (a/\alpha)^2} \qquad (5.26)$$

As $\alpha \to \infty$, the concentration becomes uniform over the sphere and eqn (5.26) becomes eqn (5.20). Equation (5.26) is of the same type as eqn (5.20), with a sum of exponential decays of which the first term becomes dominant at longer times, i.e. $\ln(m/m_0)$ becomes linear with respect to time at longer times. However, for finite values of α/a, and especially for values less than unity, the higher terms in the sum of exponentials fall off less rapidly in importance than was the case for uniform initial distribution. This results in the initial rapid extraction being of greater importance and a lowering of the intercept of the linear portion.

The value of the intercept, $-I$, may be found from the natural logarithm of the coefficient of the first term in the series ($n = 1$), i.e.

$$-I = \ln \left[\frac{2(a/\alpha)^3}{(a/\alpha)^2 - 2(a/\alpha) + 2[1 - \exp(-a/\alpha)]} \cdot \frac{1}{\pi^2 + (a/\alpha)^2} \right] \qquad (5.27)$$

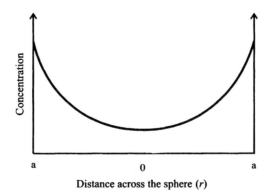

Fig. 5.13 Model concentration profile across a sphere.

Table 5.3 Values for the intercept, I, for extractions from both a sphere and a film with non-uniform initial solute distribution for various values of the ratio of the distance parameter for the distribution, α, to the sphere radius, a, or the thickness of the film, L	α/L or α/a	I(sphere)	I(film)
	∞	0.4977	0.2100
	0.3	0.5906	0.3820
	0.1	1.5051	1.0103
	0.05	2.2271	1.6338
	0.03	2.7623	2.1291
	0.01	3.8930	3.2199

Some values of I for various values of α/a are given in Table 5.3. and it can be seen that I becomes larger as α/a becomes smaller and the initial concentration distribution becomes more non-uniform. The model is not ideal as it assumes that the diffusion coefficient is constant over the sphere, which may not be the case for, say, plant and soil matrices, where the centres of the matrix particles may be comparatively impenetrable. However, it does explain qualitatively why intercepts are larger than their theoretical values for uniform concentration.

5.4.3 Limiting equations for a film

Here the form of the extraction curve for a film is considered, when the compound being extracted is very soluble in the supercritical fluid and the partition coefficient, K, is high. In this case as $ha \rightarrow \infty$, $\lambda n \rightarrow (2n-1)\pi$, as given in eqn (5.19), and eqn (5.16) becomes

$$m/m_0 = \frac{8}{\pi^2} \sum_{n=0}^{\infty} \frac{1}{(2n + 1)^2} \exp[-(2n + 1)^2 t/t_c] \qquad (5.28)$$

where t_c is now given by eqn (5.17). Again eqn (5.28) can be obtained more directly by solving the diffusion equation within the film after replacing the surface boundary conditions by

$$c(0, t) = c(L, t) = 0 \qquad (5.29)$$

A plot of eqn (5.28) is similar to the corresponding plot for a sphere, shown in Fig. 5.11. However, $\ln(8/\pi^2)$ is equal to -0.2100, and so the intercept of the linear portion at longer times, extrapolated back to the $t=0$ axis, is smaller. The linear portion becomes dominant more rapidly and the fall to the linear portion is smaller. This is a feature of the lower surface to volume ratio $1/L$ for the infinite slab (neglecting the edges), as compared with $3/a$ for the sphere.

Figure 5.14 shows some experimental results for extraction from polymer film (Bartle *et al.* 1991). The sample was a film of poly(ethylene terephthalate)

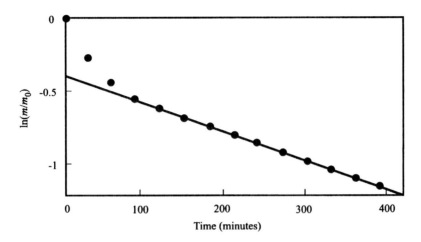

Fig. 5.14 Extraction of the cyclic trimer from poly(ethylene terephthalate) with CO_2 at 70 °C and 400 bar.

(PET), 1.2 mm in thickness, extraction was carried out at 70 °C with CO_2 at 400 bar, and results are shown for the extraction of the cyclic trimer of ethylene terephthalate. Figure 5.14 is a curve of the expected form with a steeper portion falling to a straight line after approximately 125 minutes. However, the value of I at 0.39 is above the theoretical value of 0.21 (showing similarity to the results obtained in the studies using the spherical model). This can be explained by the use of a non-uniform initial concentration distribution, as was done for a sphere in the last section, and will be discussed for a film in the next section. However, it is likely that the distribution of the cyclic trimer is uniform in the film. A plausible explanation, in this case, is that a higher proportion of the oligomer near the surface is extractable under the conditions used. (It should be mentioned that the amount of cyclic trimer extractable under these conditions is considerably below that obtained by more rigorous extraction methods: an example of the existence of 'extractable' and 'non-extractable' material in SFE.) From the slope obtained from Fig. 5.14 and the thickness of the film, a value for the diffusion coefficient of the cyclic trimer in PET at 70 °C can be calculated to be $2.1 \times 10^{-13} \, \mathrm{m^2 \, s^{-1}}$. No literature value is available, but the result has the correct order of magnitude, by comparison with other diffusion coefficients in polymers (Mills 1986).

5.4.4 A film with non-uniform initial concentration distribution

A suitable non-uniform concentration distribution for a film is given by

$$c(x, 0) = c_0[\exp(-x/\alpha) + \exp(-(L - x)/\alpha)] \tag{5.30}$$

where α is a distance parameter as before. The concentration is therefore falling off exponentially from both surfaces of the film. Following the procedures of Section 5.4.2 gives the following equation for the recovery curve:

$$m/m_0 = \frac{4\alpha}{L} \cdot \frac{1 + \exp(-L/\alpha)}{1 - \exp(-L/\alpha)} \sum_{n=0}^{\infty} \frac{\exp[-(2n+1)^2 t/t_c]}{1 + ((2n+1)\alpha\pi/L)^2} \qquad (5.31)$$

As before, the value of the intercept, $-I$, may be found from the natural logarithm of the coefficient of the first term in the series ($n = 0$), i.e.

$$-I = \ln\left[\frac{4\alpha}{L} \cdot \frac{1 + \exp(-L/\alpha)}{1 - \exp(-L/\alpha)} \cdot \frac{1}{1 + (\alpha\pi/L)^2}\right] \qquad (5.32)$$

Some values of I for various values of α/L are given in Table 5.3, showing the same trend of rising I as the concentration becomes more non-uniform as was found for a sphere.

5.5   Control by removal in solution only

5.5.1   Limiting equations

When ha is very small, corresponding to poor partition into the fluid and rapid diffusion, SFE behaves exponentially and the plot of $\ln(m/m_0)$ versus time becomes a straight line. The curve for $ha = 1$ in Fig. 5.4 can be seen to be close to this condition. For $ha \to 0$, therefore, only the first term in the sum in eqn (5.8) is significant. Equation (5.10) gives a value of $3ha$ for $(\lambda_1)^2$ and using this value eqn (5.8) for a sphere becomes

$$m/m_0 = \exp\left[-\frac{3hat}{\pi^2 t_c}\right] \qquad (5.33)$$

Using eqn (5.5) for h, eqn (5.9) for t_c, and eqn (5.14) for the relationship between the total volume, V, and surface area, A, of all the spheres, eqn (5.33) can be transformed into

$$m/m_0 = \exp\left[-\frac{KFt}{V}\right] \qquad (5.34)$$

In this situation, only partition is important in controlling extraction, which is first order, with the rate coefficient being determined by the product of the partition coefficient and the ratio of the volume flow rate of the fluid to the volume of the matrix being extracted. It can similarly be shown that the recovery equation for a film under the influence of both transport and removal by solvation, eqn (5.14), also becomes eqn (5.34) in the limit as $ha \to 0$. This is expected because, when only removal by solvation is rate determining, the geometry of the matrix is immaterial.

If V' is the volume of fluid passed at a particular time, eqn (5.34) can be rewritten as

$$m/m_0 = \exp\left[-\frac{KV'}{V}\right] \qquad (5.35)$$

Thus, when removal by solvation is the only significant step, the amount of solute removed depends only on the volume of fluid passed compared with the volume of the matrix. This result is a feature of the fact that the partition coefficient, K, is defined as a ratio of two concentrations in terms of amounts per unit volume. If a different partition coefficient, K, is used, defined in terms of a ratio of mass fraction in the fluid and matrix, eqn (5.35) can be written as

$$m/m_0 = \exp\left[-\frac{KM'}{M}\right] \qquad (5.36)$$

where M is the mass of the matrix and M' is the mass of fluid passed. A similar equation can also be written in terms of moles and mole fractions, as will be done below.

Figure 5.15 shows some results for the extraction of lycopene from tomato paste, dried by mixing with diatomaceous earth, using samples of 0.5 g and

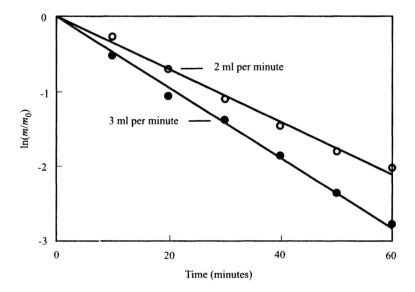

Fig. 5.15   Extraction of lycopene from tomato paste with CO_2 at 100 °C and 400 bar at two flow rates.

two flow rates, measured as liquid at the pump. The recoveries are seen to be exponential from the straight-line logarithmic plots passing through the origin. The faster flow rate gives faster recovery, although the ratio of the slopes is not exactly 2 : 3, as would be predicted from eqn (5.34). However, plots of percentage recovery versus the amount of CO_2 passed, shown in Fig 5.16, roughly coincide for the two flow rates as predicted from eqn.(5.35).

5.5.2 Extraction of an unbound solute

In some cases, the compound or compounds to be extracted are present in large quantities and appear to act independently of the matrix. They seem not to be bound to the matrix material and their extraction is governed not by a partition coefficient, but by their solubility in the fluid. Curves of percentage recovery versus time for a given flow rate are linear and the following equation applies

$$m/m_0 = 1 - FSt \qquad (5.37)$$

where S is the solubility of the solute in the fluid in terms of mass per unit volume. The situation is similar to the determination of solubility by the 'extraction' of a pure solute. Figure 5.17 shows an example of this which is the extraction of fat from crisps or potato chips. Extraction was carried out with pure carbon dioxide at $60\,°C$ and 335 bar (Hawthorne *et al.* 1995). The recovery curve is linear except near 100 per cent, where it tails off showing some

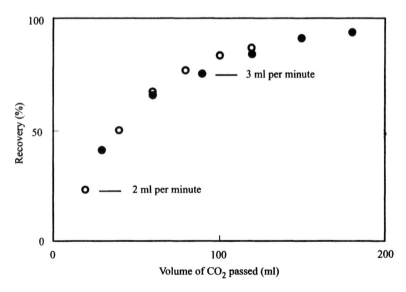

Fig. 5.16 Extraction of lycopene from tomato paste with CO_2 at $100\,°C$ and 400 bar at two flow rates plotted against the amount of CO_2 passed.

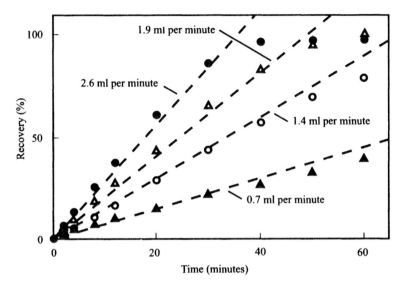

Fig. 5.17 Extraction of fat from potato chips with CO_2 at 60 °C and 330 bar at various flow rates (Hawthorne *et al.* 1995), redrawn with permission from the authors.

interaction with the matrix. The dashed lines in Fig. 5.17 have slopes proportional to the flow rate, showing that the results are consistent with eqn (5.37) below 80 per cent recovery. A solubility for the fat, obtained from the slopes of these lines is 0.01 in terms of mass fraction in the fluid, which is consistent with measured values for triglycerides (Bartle *et al.* 1991*a*), bearing in mind that the fat is an unknown mixture.

5.5.3 Separation of two liquid substances

A feature of SFE is that extraction can be selective to some extent and in this section a particular situation is examined of the separation of two liquids under conditions where removal by solution is the rate-determining step in extraction. To ensure this is the case, the liquid is effectively contacted with the fluid by rapid stirring or by supporting it on an inert material, forming a very thin surface coating, so that diffusion through the coating is fast. The separation depends on the solubility of the first substance being larger than that of the second. Then during dynamic SFE the majority of first substance is extracted with a small amount of the second, leaving behind the majority of the second substance, with a small amount of the first. Separation is not therefore complete and is analogous to a single-stage distillation. Conversion into a multi-stage process is discussed in the next chapter.

Modelling of the process is simplified if two assumptions are made. Firstly, it is assumed that the fluid substance does not dissolve in the mixture. This will not be true in general, but the effect will not qualitatively change the modelling behaviour. Secondly, the mixture is assumed to be ideal and to behave as an ideal mixture, so that there is a relationship $y_i/x_i = S_i$ between the mole fractions of the components in the mixture, x_i, and those in the fluid, y_i, with S_i being the solubility for a particular pressure and temperature, in terms of mole fraction. The rate of removal of each component is equal to the molar flow rate, F_x, times the mole fraction of the component in the fluid:

$$\mathrm{d}n_i/\mathrm{d}t = -F_x y_i = -F_x S_i x_i \tag{5.38}$$

where n_i is the number of moles of component i (with n the total number of moles) at any particular time. To simplify the algebra, a quantity $a_i = F_x S_i$ is defined and thus

$$\mathrm{d}\ln n_i = -a_i/n\,\mathrm{d}t \tag{5.39}$$

$$\ln n_i = -a_i \int_0^t \mathrm{d}u/n + C \tag{5.40}$$

where u is a dummy variable. At $t = 0$, n_i is given the value n_i° and thus $C = \ln n_i^\circ$ and

$$\ln n_i = \ln n_i^\circ - a_i \int_0^t \mathrm{d}u/n \tag{5.41}$$

This equation gives the variation of n_i with t. However, calculations cannot be easily carried out with this equation and it is better to define a quantity

$$z = \int_0^t \mathrm{d}u/n \tag{5.42}$$

and to calculate both t and mole fraction variation from z. In terms of z

$$\ln n_i = \ln n_i^\circ - a_i z \tag{5.43}$$

and thus n_i can be calculated from z using

$$n_i = n_i^\circ \exp(-a_i z) \tag{5.44}$$

To obtain the equation for the calculation of t in terms of z, eqn (5.44) is summed over both compounds and then divided by n to obtain

$$1 = \sum_i n_i^o/n \exp[-a_i \int_0^t (du/n)] \qquad (5.45)$$

Multiplying both sides by du and integrating between 0 and t gives

$$t = -\sum_i (n_i^o/a_i \exp[-a_i \int_0^t (du/n)] + D \qquad (5.46)$$

At $t = 0$, we find $D = \sum_i (n_i^o/a_i)$ and thus

$$t = \sum_i (n_i^o/a_i)[1 - \exp(-a_i z)] \qquad (5.47)$$

Thus eqns (5.44) and (5.47) enable n_i and t to be calculated from a variable z allowing n_i to be obtained indirectly as a function of t.

A particular situation is now considered, where it is desired to purify component 2 by separating it from component 1, which is 10 times more soluble than component 2. This can be done by extracting component 1 from the mixture and stopping the extraction when the majority of component 1 has been removed. The mole fractions of each component remaining, n_i, are calculated from eqns (5.44) and (5.47) above. For model parameters $n_1^o = n_2^o = 1$, $a_1 = 1$ and $a_2 = 0.1$ are chosen, and after calculation the results shown in Fig. 5.18 are obtained. Figure 5.18(a) shows the amounts of each component remaining as a function of the arbitrary time units. As expected, the amount of component 1 remaining, n_1, falls more rapidly than that for component 2, n_2, removing it from the mixture. Figure 5.18(b) shows the mole fraction of component 2 remaining, which starts at the initial value of 0.5 and approaches unity at long times. As can be seen, 97 mol per cent of 2 can be obtained after 4 time units, but by then some 30 mol per cent of 2 will have been lost. If a purity of 80 mol per cent of 2 is acceptable, then this could be obtained after 2 time units with only some 10 per cent of 2 lost. Because the molar mass of component 2 will typically be much higher than that of 1, a purity of 80 mol per cent will be higher in terms of mass per cent. If the molar mass of component 2 is twice that of component 1, 80 mol per cent purity will correspond to 89 mass per cent purity.

5.5.4 Fractionation of polymers

As an example of the separation of more complex mixtures, the fractionation of polymers is now discussed. Treatment of the physical chemistry of separation is carried out in the next two sections, and this is preceded in this section by

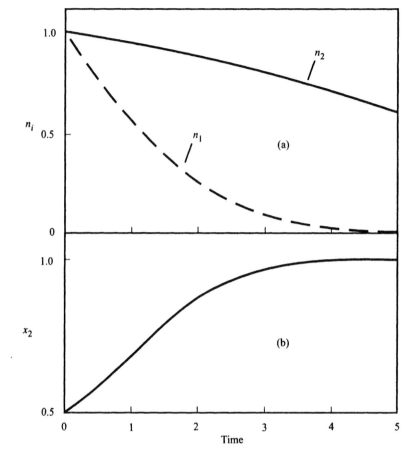

Fig. 5.18 Modelling of the extraction of a binary liquid mixture.

outlining the principles of two of the most important practical procedures used
to effect fractionation. These are, to use the nomenclature of McHugh and
Krukonis (1994), 'isothermal decreasing pressure profiling' and 'isothermal
increasing pressure profiling'. The former is carried out using an extraction
system, such as that depicted in Fig. 5.1, except that it now contains a number
of collection vessels or separators, each operating at a lower pressure than the
previous one in the train. This is illustrated schematically in Fig. 5.19, where
successively decreasing pressures are maintained by a series of control valves.
As can be seen, the process can be operated in a continuous fashion. The
polymer feed is extracted in one vessel under conditions where most of it
dissolves, leaving only the heavier components, which are eluted from the
vessel as fraction 1. In the first separator, the pressure is lowered to precipitate

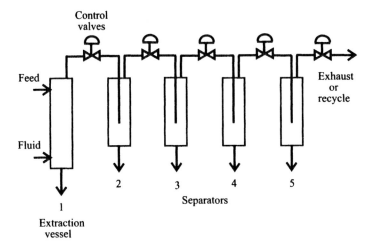

Fig. 5.19 Schematic diagram of a system for polymer fractionation by 'isothermal decreasing pressure profiling'.

a lighter fraction, 2, carrying even lighter components to be precipitated at lower pressures in the remaining separators to give fractions 3, 4, and 5. A patent for the fractionation of polymers by reducing the pressure was applied for as early as 1943 (Hunter and Richards 1948).

For 'isothermal increasing pressure profiling' the simpler system of Fig. 5.1 is used and can only be operated as a batch process. The polymer is extracted for a number of successive periods of time. In each period, the pressure is constant and higher than that in the previous period. Extraction is continued in each period until very little polymer is being extracted at that pressure. This is an arbitrary criterion, as in principle all the polymer will be eventually extracted if the process is continued for sufficient time.

5.5.5 Fractionation of a polymer at constant pressure

Time is thus an important parameter in the extraction of a polymer by a supercritical fluid, and the equations developed in Section 5.5.3 can be used to model the process as a function of time. The equations were developed for a binary mixture, but in fact apply also to a multicomponent mixture, such as a polymer. The polymer is considered to contain n_i moles of the ith oligomer with n moles in total, with initial values of n_i° and n°. F_x is the flow rate of fluid in moles per unit time through the extraction cell and the parameter a_i for the ith oligomer is defined by $a_i = F_x S_i$, where S_i is the solubility of the ith oligomer. With these redefinitions, eqns (5.42) and (5.43) apply. It is also necessary to have information about the solubility of a polymer of a particular chain length. For the present purpose a model is used (Fjeldsted *et al.* 1983), which is simple and yet is successful in the qualitative prediction of polymer extraction behaviour and also in the understanding of the supercritical fluid chromatography of polymers, as will be described in Section 7.3.1. At constant pressure and temperature, this model assumes that the solubility of each oligomer is a constant fraction of that of the previous oligomer, i.e. the solubilities of the oligomers fall off exponentially with chain length.

A calculation is now carried out for a simplified model polymer of 10 oligomers of equal amounts, i.e. $n_i^o = 1$ for all i and $n = 10$. The flow rate of the fluid will be assumed to be constant and it will be also assumed that $a_i = 1/2^{i-1}$ consistent with an exponential fall-off of solubility with i. The initial quantity calculated is the mole fraction of oligomer i in the polymer being extracted at a particular time (not including the fluid), z_i, which can be calculated from eqn (5.38) and the definition of a_i to be

$$z_i = (dn_i/dt)/(dn/dt) = a_i n_i / \sum_i a_i n_i \qquad (5.48)$$

The results are presented in Fig. 5.20 for the first six oligomers. Note that the mole fraction of each oligomer peaks successively and that these peaks are progressively spread out with time as the oligomers become less soluble. The result is similar to chromatography with poor resolution.

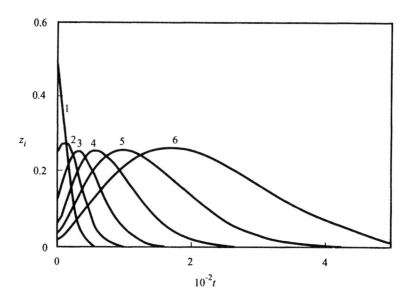

Fig. 5.20 Plots of the mole fraction of oligomer i in the polymer being extracted, z_i, using constant pressure at time t for an artificial polymer consisting of equal amounts of 10 oligomers.

The number-averaged molar mass can be written as $<M>_n = (1/n) \times \sum_i n_i i m$, where m is the mass of the monomer unit. The average molar mass emerging at any particular time is thus given by

$$<M>_n = [1/(dn/dt)] \sum_i (dn_i/dt) im \qquad (5.49)$$

From eqn (5.38) and the definition of a_i this becomes after cancellation of n

$$<M>_n /m = \sum_i a_i n_i i / \sum_i a_i n_i \qquad (5.50)$$

Calculations of $<M_n>/m$, from the same simplified model data are plotted in Fig. 5.21. The number-averaged molar mass rises with time as the lower oligomers are exhausted, but this rise slows, due to the low solubility of the higher oligomers. This is what happens in each step of 'isothermal increasing pressure programming', where the rise in molar mass and the amount being extracted tails off. A new extraction with similar behaviour is then obtained by stepping up the pressure.

The polydispersity, P, is defined in general as the ratio of the weight-averaged molar mass to the number-averaged molar mass and can be represented in terms of the variables used here as $P = n\sum_i n_i i^2/(\sum_i n_i i)^2$. The polydispersity of the extracted polymer emerging at any particular

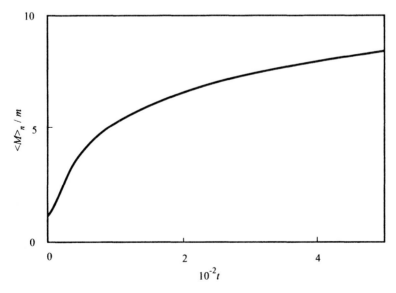

Fig. 5.21 A plot of $<M>_n/m$, the number-averaged molar mass, versus time, t, using the same artificial data as in Fig. 5.20.

time is therefore given by

$$P = \left[\sum_i (\mathrm{d}n_i/\mathrm{d}t)\right]\left[\sum_i (\mathrm{d}n_i/\mathrm{d}t)i^2\right] \bigg/ \left[\sum_i (\mathrm{d}n_i/\mathrm{d}t)i\right]^2 \qquad (5.51)$$

From eqn (5.38) and the definition of ai this becomes after cancellation of n

$$P = \left[\sum_i a_i n_i\right]\left[\sum_i a_i n_i i^2\right] \bigg/ \left[\sum_i a_i n_i i\right]^2 \qquad (5.52)$$

A plot of P calculated from the simplified model system is given in Fig. 5.22. The polydispersity is seen to fall with time, as the higher less soluble oligomers are being extracted slowly.

5.5.6   Fractionation of a polymer using a density program

For SFE at constant pressure, excessively long times are needed towards the end of the separation, as can be seen from Fig. 5.20, where successive peaks emerge more slowly. In the methods known as 'isothermal increasing pressure programming', the pressure is increased in steps each time the extraction slows to obtain a fraction of higher average molar mass. If time is to be included in the modelling of polymer fractionation, the pressure changes need to be described as a function of time, i.e. be carried out according to a programme. This

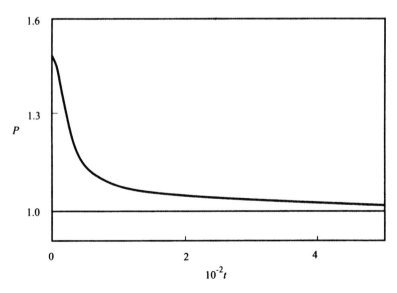

Fig. 5.22   A plot of the polydispersity, P, versus time, t, using the same artificial data as in Fig. 5.20.

approach has been used in supercritical fluid chromatography to obtain chromatograms with suitably distributed peaks. For the chromatography of polymers, for example, it is desirable to have the oligomer peaks equally spaced with respect to time and for this an asymptotic density programme was developed (Fjeldsted *et al.* 1983), as will be described in Section 7.3.1. In the case of polymer fractionation by SFE, the requirements may be different and a likely situation is that constant polydispersity is required. At constant poly-dispersity, a greater range of oligomers are present at higher average molar masses. Thus for constant polydispersity, successive peaks needs to emerge more rapidly as the process continues. This effect occurs in supercritical fluid chromatography with a linear density gradient and this programme is found to give approximately constant polydispersity in extraction and will be used here.

The solubility behaviour used by Fjeldsted *et al.* (1983), which was successful for developing density programming for supercritical fluid chromatography, is used here. They assumed that the solubility of oligomers is given by

$$S_i = b_1 \exp[-(b_2 - b_3\rho)i] \tag{5.53}$$

where b_i are constants for a particular polymer and fluid. If a linear density program of $\rho = b_4 + b_5 t$ is applied, where b_4 and b_5 are similar constants, the parameter a_i defined earlier, is given by

$$a_i = F_x b_1 \exp[-(b_2 - b_3 b_4 - b_3 b_5 t)i] \equiv c_1 \exp[-c_2(1 - c_3 t)i] \tag{5.54}$$

where the c_i are constants for a particular polymer, fluid, and density pro-gramme. The solubility behaviour used in the last section was a simple example of eqn (5.54), with $c_1 = 2$, $c_2 = 0.693$, and $c_3 = 0$.

Now that a_i is a function of t, the algebraic analysis used in Section 5.5.3, in particular the integration of eqn (5.39) to give eqn (5.40), is no longer valid. The calculations are therefore carried out numerically. A more realistic model of a low polymer is now used in which there are 100 oligomers with amounts distributed according to a gaussian distribution of $n_i^0 = \exp[-(i - 50)^2/1250]$ a_i was taken to be given by eqn (5.54) with $c_1 = 2$, $c_2 = 0.693$, and $c_3 = 10^{-5}$. Figure 5.23 shows the quantity z_i, the mole fraction of a particular oligomer in the extract, for oligomers 10–15 and thereafter every tenth oligomer. As can be seen, the oligomers emerge in sequence, but more and more rapidly with time, in contrast to the situation illustrated by Fig. 5.20.

The number-averaged molar mass and polydispersity from the same model polymer and density programme were also calculated. The solid line in Fig. 5.24 shows the natural logarithm of the average molar mass, which can be seen to be roughly linear, with a slight 'S' shape visible at the ends, where little material is being extracted. The solid line in Fig. 5.25 shows the polydispersity, which is approximately constant over most of the range, although it peaks slightly in the

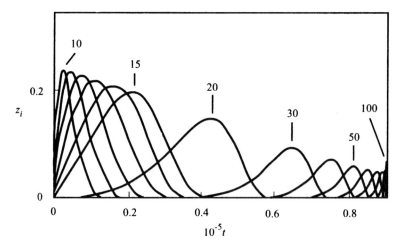

Fig. 5.23 Plots of the mole fraction of oligomer i in a polymer being extracted, z_i, using linear density programming at time t for oligomers 10–15 and then every tenth oligomer from an artificial polymer consisting of 100 oligomers with a gaussian distribution.

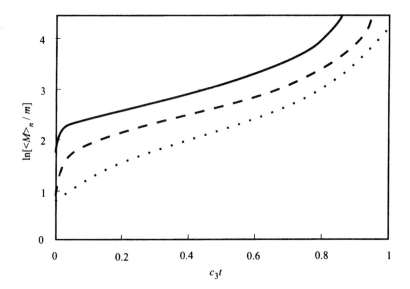

Fig. 5.24 Plots of the natural logarithm of $<M>_n/m$, the number average molecular mass of extracted polymer, versus time, t, from the same model polymer as in Fig. 5.23, using density programmes such that $a_i = 2\exp[-0.693(1 - c_3 t)i]$ and: $c_3 = 10^{-5}$ (solid line); $c_3 = 10^{-4}$ (dashed line); $c_3 = 10^{-3}$ (dotted line).

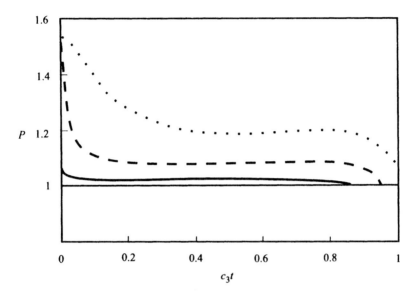

Fig. 5.25 Plots of the polydispersities, P, of the extracted polymer versus time, t, from the same model polymer as in Fig. 5.23, using density programmes such that $a_i = 2\exp[-0.693(1 - c_3t)i]$ and: $c_3 = 10^{-5}$ (solid line); $c_3 = 10^{-4}$ (dashed line); $c_3 = 10^{-3}$ (dotted line).

middle. Calculations were also carried out for the same model polymer with more rapid density programmes, in which c_1 and c_2 were given the same values as before, but in which $c_3 = 10^{-4}$ (dashed lines in Figs 5.24 and 5.25) and 10^{-3} (dotted lines in the figures). More rapid density programmes are seen to give similar shaped curves, but with a lower molar mass curve and a higher polydispersity.

5.5.7 Comparison with experimental results for SFE of polymers with linear density programming

Some experimental results are now presented, which show qualitative agreement with the modelling described in the last section (Clifford *et al.* 1997a). Polyisobutene is not very soluble in carbon dioxide and only low oligomers can be extracted with it. However, this also means that high polydispersities can be obtained. The starting material had a number-averaged molar mass of 1274 and a polydispersity of 1.54. A preliminary extraction was carried out at 50 °C and 636 kg m^{-3} for 10 minutes to remove the tail at low molar masses and to produce the first fraction. Extraction was then carried out at 50 °C using a linear density programme rising from 656 kg m^{-3} to 1011 kg m^{-3} at a rate of 6.32 kg m^{-3} per minute. The flow rate of carbon dioxide, calculated as liquid carbon dioxide at the pump, was 10 ml per minute. Eight further fractions were collected at equal density intervals of $45\,kg\,m^{-3}$. 52 per cent by mass of the polymer was extracted. The number-averaged molar masses of the fractions were between 342 and 1528, and are plotted as logarithms in Fig. 5.26 against density, which is equivalent to time as the density programme is linear. The shape of the curve is similar to the first part of the curves in Fig. 5.24, bearing in mind that only half of the polymer was extracted. The polydispersity, shown in Fig. 5.27, falls and then rises slightly, as in the first part of the theoretical curves of Fig. 5.25. These results show agreement between experiment and prediction and also that, with the right density programme low and approximately constant polydispersities can be obtained, except at the beginning.

Results for the SFE of dimethylsiloxane are now presented to demonstrate the effect of density ramp speed on the molar mass and polydispersity, and to show that this agrees with the theoretical predictions of Figs 5.24 and 5.25. The large changes in the rate of rise of density, possible in calculations, are impracticable and a ratio of about 2 was used in the experiments described (Clifford *et al.* 1997*a*). Polydimethylsiloxane is much more soluble in carbon dioxide than polyisobutene and the fractions obtained had higher poly-dispersities. The starting material had a number-averaged molar mass, meas-ured by the procedures described above, of 127 682 and a polydispersity of 1.32.

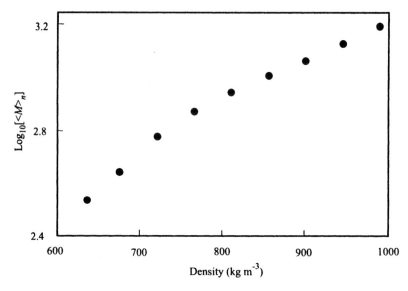

Fig. 5.26 SFE of polyisobutene using a linear density programme, showing the logarithm of the number-averaged molar masses of the fractions versus density.

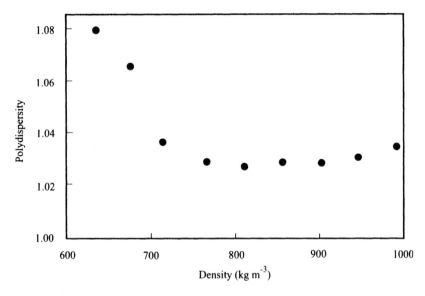

Fig. 5.27 SFE of polyisobutene using a linear density programme, showing the polydispersity of the fractions versus density.

Both extractions described were carried out at 40 °C and with a flow rate of carbon dioxide, calculated as liquid carbon dioxide at the pump, of 6.5 ml per minute. In each case a preliminary extraction was carried out at low densities, to remove the tail at low molar masses, and discarded.

Two linear density programmes are reported, differing in the rate of rise of density by a factor of 2.13. For the lower speed ramp (filled circles in Figs. 5.28 and 5.29) the density rose from $889 \, kg \, m^{-3}$ to $1014 \, kg \, m^{-3}$ at a rate of $0.94 \, kg \, m^{-3}$ per minute and seven fractions were collected at equal density intervals of $18 \, kg \, m^{-3}$. For the higher-speed ramp (open circles in Figs. 5.28 and 5.29) the density rose from $912 \, kg \, m^{-3}$ to $1022 \, kg \, m^{-3}$ at a rate of $2.00 \, kg \, m^{-3}$ per minute and seven fractions were collected at equal density intervals of $16 \, kg \, m^{-3}$. 86 and 84 per cent by mass of the polymer was extracted in the two experiments, respectively. The number-averaged molar masses of the fractions for both runs are plotted as logarithms in Fig. 5.28 against density, which is equivalent to time as the density programme is linear. The filled circles for the lower-speed ramp lie above those for the higher-speed ramp, as predicted in Fig. 5.24. The polydispersities, shown in Fig. 5.29, have the filled circles for the lower-speed ramp below those of the higher-speed ramp, as predicted by Fig. 5.25. Thus the effect of ramp speed has the same qualitative effect on the experimental results as it does on the theoretical predictions.

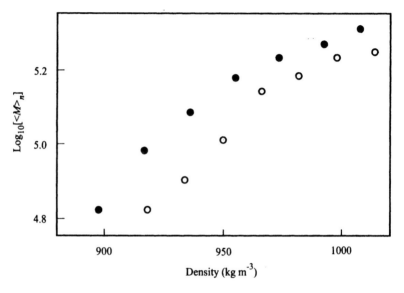

Fig. 5.28 SFE of polydimethyisiloxane using a linear density programme, showing the logarithm of the number-averaged molar masses of the fractions versus density, with filled circles for a lower rate of density rise and open circles for a higher rate.

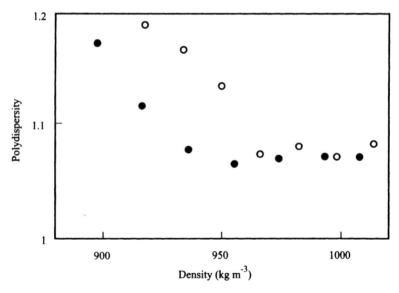

Fig. 5.29 SFE of polydimethylsiloxane using a linear density programme, showing the polydispersity of the fractions versus density with filled circles for a lower rate of density rise and open circles for a higher rate.

5.6 Control by reversible release, transport through the matrix, and removal by solution

5.6.1 The matrix effect

Of the three factors which are thought to control SFE, that of matrix effects is the least well understood. The most important characteristic of the matrix effect is that extraction, which is initially successful, becomes very slow before 100 percent recovery. Often the extraction is so slow that the amount being extracted is below detection limits and extraction appears to have stopped. If 100 μg of chrysene is doped on to 1 g of untreated silica gel, 99 per cent is recovered in about 30 minutes using CO_2 at 50 °C and 300 bar (Walker 1995). If the silica gel is previously heat-treated to 400 °C, recoveries of 72 per cent are obtained with the same conditions after 60 minutes. Towards the end recoveries are below detection limits. The recovery curve is shown in Fig. 5.30. Even if the heat treatment temperature is reduced to 200 °C, only 69 ± 4 per cent is recovered using carbon dioxide at 50 °C and 200 bar in 60 minutes. An initial conclusion is that heat treatment of the silica produces some very active adsorption sites, which fix 30 μg of the chrysene to the surface. If that is the case, spiking by a smaller amount on to the silica in the same volume of solvent should produce much lower recoveries. If 50 μg is spiked on, only 20 μg should be recovered or 40 per cent and if 25 μg is spiked on, none should be recovered. In fact the recoveries at the same conditions of 50 °C and 200 bar

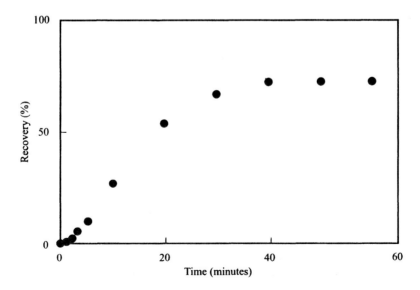

Fig. 5.30 SFE of chrysene spiked on to silica, heat treated at 400 °C, with CO_2 at 50 °C and 400 bar.

after 60 minutes are 56 ± 2 per cent and 42 ± 7 per cent, respectively. Thus a model involving two types of adsorption site of different activities is too simplistic.

The example just described is extreme in that extraction appears to have completely stopped below 100 per cent. Many extractions however, which appear to be approaching a final recovery of less than 100 per cent, are still, in fact, slowly rising. This can be demonstrated by carrying out extractions for an abnormally long period. Some of the results for the extraction of polyaromatic hydrocarbons from contaminated soil are shown in Fig. 5.31. SFE was carried out with pure CO_2 at 55 °C and 400 bar and a flow rate of 0.9 ml per minute (Clifford *et al.* 1995). 100 per cent recovery is based on the sum of two extractions plus the amount recovered by 14 hours of ultrasonication of the SFE residue in methylene chloride. These curves show the following features. Firstly, there is an initial slower extraction at very short times. This is not very obvious from all the results as the initial extraction period is outside the timescale for this effect, but it is quite visible for the extraction of indeno[1,2,3-*cd*]pyrene from contaminated soil and some other curves show vestiges of this effect. This cannot be due to experimental start-up effects, as these would be the same for all compounds. The effect is also shown by the extraction of chrysene from silica, shown in Fig. 5.30. There then follows a

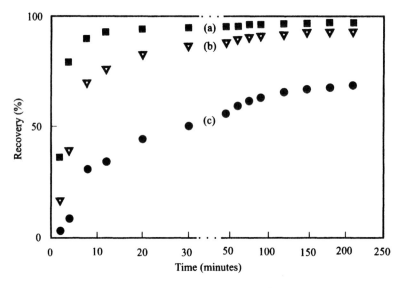

Fig. 5.31   SFE of (a) chrysene (b) benzo[*b*]fluoranthene plus benzo[*k*]fluoranthene and (c) indeno[1,2,3-*cd*]pyrene from contaminated soil using CO_2 at 400 bar and 55 °C over an extended period, shown with two different time scales.

more rapid extraction phase which ceases often well below 100 per cent recovery. Finally there is a much slower final extraction phase heading towards 100 per cent recovery.

5.6.2 Modelling of extraction

SFE involving matrix effects is essentially complex and any model will need to neglect some possible features. It was demonstrated in the last section that an approach involving two types of adsorption site is too simplistic, and there is likely to be a range of interaction energies between the matrix and the compound to be extracted. Here a simple model is presented, which involves only one type of solute–matrix interaction which interacts with the diffusion and solvation processes previously discussed. It is an extension of the models described previously, and qualitatively describes matrix effects. It is not claimed, however, to be more than a superficial description of SFE involving matrix effects. All of the steps (2), (3), and (4) of the model described in Section 5.2 and illustrated in Fig. 5.3 are now used in the model. The new step to be considered is that of reversible release from a bound state, which may arise by adsorption, location behind a cell wall, or restraint by interwoven polymer chains. For convenience the reversible release process will be described as absorption and desorption in the following discussion, although other possible processes can be treated in the same way. It will be described by the first-order rate coefficients k_1 (e.g. desorption) and k_2 (e.g. readsorption).

The matrix is considered here to consist of spherical matrix particles of radius a, and any point in a particle is described by its distance, r, from the centre. We consider concentrations of the solute in both the adsorbed and free (or bound and unbound by another mechanism) states, c' and c respectively, which are a function of r and t. It is assumed that the solute is totally adsorbed initially and that its concentration is uniform throughout the particle at c_0. This is expressed by boundary conditions:

$$c'(r,0) = c_0; \qquad c(r,0) = 0 \qquad (5.55)$$

As removal by solvation is being taken into account, the boundary condition expressed by eqn (5.4), with h defined by eqn (5.5) also applies. The equations for the rates of change of the two concentrations are

$$\partial c'/\partial t = -k_1 c' + k_2 c \qquad (5.56)$$

$$\partial c/\partial t = +k_1 c' - k_2 c + D\{(\partial^2 c/\partial r^2) + (2/r)(\partial c/\partial r)\} \qquad (5.57)$$

Both equations contain terms for the rates of adsorption and desorption, and eqn (5.57) also contains a term for the diffusion of the desorbed solute. The differential eqns (5.56) and (5.57) are now solved in conjunction with the boundary conditions; eqns (5.4) and (5.55). This can be done by the method of Laplace transforms. After integration over the particle, an expression for the fraction of solute remaining after time t is found to be

$$m/m_0 = \sum_{n=1}^{\infty} C_n[\exp(-p_nt)/p_n - \exp(-q_{ni}t)/q_n] \qquad (5.58)$$

The equations for the rate coefficients for these decays are

$$p_n = k_1 + k_2 + D\lambda_n^2/a^2 - [(k_1 + k_2 + D\lambda_n^2/a^2)^2 - 4k_1D\lambda_n^2/a^2]^{1/2} \qquad (5.59)$$

and

$$q_n = k_1 + k_2 + D\lambda_n^2/a^2 + [(k_1 + k_2 + D\lambda_n^2/a^2)^2 - 4k_1D\lambda_n^2/a^2]^{1/2} \qquad (5.60)$$

where λ_n are the roots of eqn (5.10) as before. The coefficients C_n in eqn (5.58) are given by

$$C_n = 6k_1h^2D\{[\lambda_n^2 + ha(ha - 1)][(k_1 + k_2 + D\lambda_n^2/a^2)^2 - 4k_1D\lambda_n^2/a^2]^{1/2}\}^{-1} \qquad (5.61)$$

Equation (5.58) and its subsidiary eqns (5.10) and (5.59) to (5.61), give a prediction of the fraction remaining in the matrix after extraction for a given time t in terms of the input parameters to the model; the rate coefficients k_1, k_2, and D/a^2, which are in units of inverse time, and the dimensionless parameter ha, which is proportional to the solubility. Equation (5.58) contains two infinite series of exponential decays, one of which is negative. The negative series is relatively fast; inspection of eqns (5.59) and (5.60) shows that q_n must lie between $k_1 + k_2$ and infinity, whereas p_n must lie between zero and k_1. Figure 5.32 is a diagram showing the first few members of each series for some representative input parameters: $k_1 = 1$, $k_2 = 1$, $(D/a^2) = 0.01$, and $ha = 1$, the units of all input and output rate coefficients being the same.

If the input parameters are such that p_1 is much smaller than k_1, it can be shown that p_1 is approximately given by

$$p_1 = 3ha[k_1/(k_1 + k_2)](D/a^2) \qquad (5.62)$$

Thus if SFE is being carried out under adverse conditions such that partition is unfavourable and thus ha is small, the rate coefficient for readsorption is greater than that for desorption, and the transport rate D/a^2 is small, all these factors combine in reducing the rate of extraction at long times. Furthermore, as discussed previously, a large fraction of the solute will extract at this rate. The quantities multiplying the exponential decays are C_n/p_n and C_n/q_n, respectively, which means that the more rapidly decaying terms in q_n are

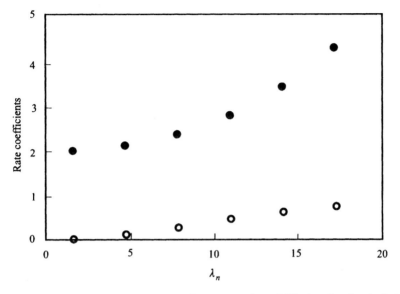

Fig. 5.32 Values for the rate coefficients for the first six terms of eqn (5.58), plotted against the first six values of λ_n, which are roots of equation (5.10), p_n (open circles) and q_n (filled circles), with the input parameters for the calculation: $k_1 = 1$, $k_2 = 1$, $D/a^2 = 0.01$ and $ha = 1$, and the units of all input and output rate coefficients the same.

numerically smaller than the terms in p_n, and that the initial rate of extraction will be zero, as on differentiation the pairs of terms will cancel as t tends to zero. C_n falls as n increases causing the series to converge. For lower values of ha (lower solubility) this convergence is more rapid, so that for compounds with poor solubility only the C_1 terms are important. Figure 5.33 shows some curves predicted from the model. The input parameters for the calculation were: $k_1 = 10$, $k_2 = 30$, $D/a^2 = 0.1$ for all curves, and $ha = 1,10$, and $\rightarrow \infty$ for curves (a), (b), and (c), respectively. The units of time are the same as in the input rate coefficients. They show the three-phase characteristics of a slow initial extraction (the rate of extraction being zero at $t = 0$ as discussed above), a more rapid rise, and then a slow-down in the final rate, as explained by eqn (5.62).

Investigation of the model equations in detail, leads to an appreciation of the physical processes occurring during the three phases of the extraction process. The separation of the process into the three phases, as described below, is only approximate and uses the schematic description of the development of concentration profiles across the spherical model particle during extraction given in Fig. 5.34. In this diagram a uniform initial distribution of solute is assumed. The three phases of extraction are described in terms of this diagram as follows. It has been assumed that initially the solute is adsorbed or

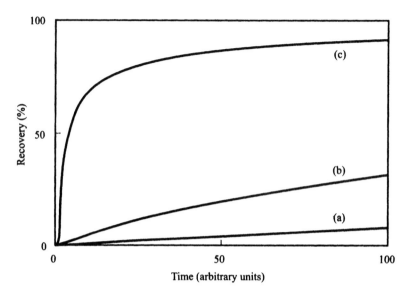

Fig. 5.33 Predicted curves of percentage recovered versus time obtained from the model which includes reversible release, transport and solvation.

otherwise held in the matrix. The initial rate of extraction is therefore zero and only builds up as the solute is released from the matrix. This initial phase is more pronounced if the rate of release, k_1, is slow. During this phase, the concentration of free solute builds up to a steady state, illustrated by the transition from situations (a) to (b) in Fig. 5.34. Thus *the initial slow phase corresponds to the attainment of a steady-state concentration of mobile solute molecules.*

Transport of the solute through the particle will only occur if there is a concentration gradient. Initially, therefore, extraction only takes place from the edge of the particles. This will erode the concentration profile at the edge of the particle, promoting transport from further inside it. The concentration profile will develop to (c) in Fig. 5.34. The rate of this process will be high, initially equal to the rate of release, k_1. *The second rapid phase corresponds to the establishment of a smooth concentration profile falling towards edge of the particle.* Once this concentration profile has been established it will decay in value, but keeping the same shape, as illustrated in the transition from (c) to (d) in Fig. 5.34. This decay will be exponential and its rate will be determined by solubility (which controls the concentration at the particle edge), diffusion, and the equilibrium between bound and free species and thus may be slow. *The final slow phase corresponds to the exponential decay of the established concentration profile.*

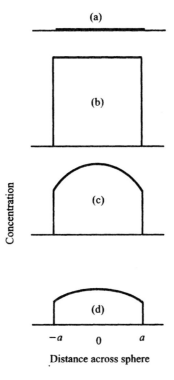

Fig. 5.34 Concentration profiles across a sphere during SFE involving matrix, transport and solvation effects.

Analogously to earlier treatments, the compound, adsorbed initially, is considered to be distributed not uniformly, but with concentration falling off exponentially from the surface, as described by eqn (5.25) and illustrated by Fig. 5.13. This equation then replaces the boundary condition for c' in eqn (5.55). The change in boundary condition has no effect on the form of eqn (5.58) or on the values of p_n and q_n. The coefficients C_n in eqn (5.58) are, however, changed and this equation becomes

$$m/m_0 = \sum_{n=1}^{\infty} C'_n[\exp(-p_n t)/p_n - \exp(-q_n t)/q_n] \qquad (5.63)$$

where C'_n is given in terms of the original C_n by

$$C'_n = \frac{C_n(\lambda_i)^3\{[(\lambda_n\alpha/a)^2 + 1][\lambda_n + \lambda_n\alpha/a(ha - 1)] + [(\lambda_n\alpha/a)^2 - 1]\lambda_n\alpha/a\}}{3ha[(\lambda_n\alpha/a)^2 + 1]\{\lambda_n^2 - 2\lambda_n^2\alpha/a + 2\lambda_n^2[1 - \exp(-a/\alpha)]\}}$$

$$(5.64)$$

As α/a tends to infinity, representing uniform distribution, eqn (5.64) becomes $C_n' = C_n$, as required. For finite values of α/a, the smaller the value, the more important are the faster decay terms in eqn (5.63). Here an exponential distribution has been considered, although equations for other distributions may be readily obtained by the same methods. The model does allow qualitative discussion of the effect of material concentrated near or away from the surface.

5.7 Extrapolation of extraction data

Extraction is never complete in finite time and to obtain the total amount of a compound obtainable by extraction extrapolation procedures can be used. These procedures may be of use both for analytical quantitation or for process design. A simple formula for carrying out such an extrapolation is now described. For all models described in this chapter, the equations approximate to an exponential decay at longer times of the form

$$m = C\exp(-t/t_c) \tag{5.65}$$

where m is the amount of solute remaining after time t, and C is a constant, for a particular extraction model.

For the extrapolation procedure described below, we require an initial extraction of an amount m_1, in a period of $t = 0{-}t_1$, by which time the extraction should have approximately reached exponential behaviour. There should then be two subsequent extractions, in equal periods of time terminating at t_2 and t_3, of amounts m_2 and m_3, respectively, i.e.

$$t_3 - t_2 = t_2 - t_1 \tag{5.66}$$

On substitution of the appropriate values at the end of each of the three extraction periods into eqn (5.65) we obtain:

$$m_0 - m_1 = C\exp(-t_1/t_c) \tag{5.67}$$

$$m_0 - m_1 - m_2 = C\exp(-t_2/t_c) \tag{5.68}$$

$$m_0 - m_1 - m_2 - m_3 = C\exp(-t_3/t_c) \tag{5.69}$$

After dividing eqn (5.67) by eqn (5.68) and also eqn (5.68) by eqn (5.69), the following two equations are obtained:

$$(m_0 - m_1)/(m_0 - m_1 - m_2) = C\exp[(t_2 - t_1)/t_c)] \tag{5.70}$$

$$(m_0 - m_1 - m_2)/(m_0 - m_1 - m_2 - m_3) = C\exp[(t_3 - t_2)/t_c)] \tag{5.71}$$

From eqn (5.66), the right-hand sides of eqns (5.70) and (5.71) are identical and so the left-hand sides can be equated. After algebraic manipulation, we obtain

$$m_0 = m_1 + \frac{m_2^2}{m_2 - m_3} \tag{5.72}$$

It can be seen that if the value of m_3 is found to be very small and can be considered zero, i.e. the extraction is almost complete after the second time period, the equation simplifies to

$$m_0 = m_1 + m_2 \tag{5.73}$$

as would be expected. If not, eqn (5.72) may be used to obtain m_0, provided the difference between m_2 and m_3 is large enough compared with the errors in the two quantities. This is not too serious a problem, as usually the last term in eqn (5.72) is small compared with m_1.

In the case of extraction from polymers, there is an advantage in working with the original (rather than ground) sample pellets, as there is a danger, suggested by some of our experiments, that the results are affected by the grinding process (perhaps by loss of solute or a change in its extractability). However, a fairly exhaustive extraction of polymer pellets of a few millimeters in diameter is likely to take 50–100 hours. The extrapolation procedure was therefore investigated for this type of system. Table 5.4 gives data for the extraction of 2,6-ditertiarybutyl-4-methylphenol (BHT) from standard poly-propylene cylinders of $\sim$3 mm in both length and diameter, with additive concentrations known to within 1 per cent w/w (Bartle *et al.* 1990*a*). Although

Table 5.4 Extrapolation to obtain final quantities in the extraction of BHT (0.2% w/w) from 178.4 mg of standard polypropylene pellets using pure CO_2 at 50 °C and at 400 bar

Extraction times (mins)	Weight extracted (μg)	Cumulative times (min)	Weight extracted (μg)
0–20	7.1		
20–60	25.0		
60–120	45.7		
120–180	36.8		
180–240	26.8	0–240	141.4
240–300	16.4		
300–360	17.1	240–360	33.5
360–480	27.8	360–480	27.8
Total	202.7		
Given total	356.8		
Total from eqn (5.72)			338.3
Difference between given total and eqn (5.72): −5.2 per cent			

extraction was carried out for 8 hours with only 57 per cent of the additive extracted, an estimate of the final amount was made using eqn (5.72) which is 5.2 per cent below the given value. From the form of the curve of $\ln(m/m_0)$ versus time for this system, and calculations from the known diffusion coefficient for BHT in high density polyethylene, it can be deduced that the linear portion of the curve is not well achieved in the first extraction period. Thus the model is assuming the tail is falling off more rapidly than is in fact the case; hence the low result. If a better result were desired, this could be obtained with the sacrifice of a longer extraction time. It appears in this case that the great majority, if not all, the additive is extractable by SFE under the conditions used, perhaps helped by the fact that the BHT molecule is a small one.

5.8 Separation of extracted solutes

After extraction, the extracted compounds must be recovered, and this is usually accomplished by lowering the pressure, as described in Section 5.1 and illustrated in Fig. 5.1. Even if the equilibrium is favourable to separation, the compounds may be precipitated as small droplets or even aerosols and trapping may be difficult. Particular engineering designs of separators have been produced to help with this. Alternatively, if the solubility of the compounds is falling with temperature at constant pressure, as is often the case with involatile solutes, separation may be achieved by raising the temperature at constant pressure. This avoids recompressing the fluid for recycling. For very volatile solutes, the degree of separation and therefore recovery by depressurization may be unsatisfactory, as will be illustrated below. In such a case, alternative methods of separation may be considered. One of these is to adsorb the solute on to a solid, such as silica or active carbon, to concentrate it, and remove it later by extraction at a higher pressure or by thermal desorption. Another is to use molecular filters, such as zeolites, which will allow the fluid but not the solute molecules to pass through, although it is at present difficult to use these on a large scale.

For an involatile solute the solubility at low pressures may be predicted by the methods of Section 3.5 in order to estimate the degree of recovery. More of a problem is experienced for volatile solutes and this is now illustrated using the separation of α-pinene from carbon dioxide as an example. Lower temperatures are optimum for the separation of volatiles and the solid line in Fig. 5.35 gives the solubility of α-pinene in carbon dioxide calculated at 40 °C and low pressures using the Peng–Robinson equation. An estimated binary interaction parameter of 0.1 was used, with the dashed lines showing schematically the behaviour from the ends of the calculated curve. The lowest point of the curve is $x_2 = 0.0011$, which corresponds to 0.34 mass per cent. Separation of this compound is therefore likely to be a problem, as the concentration

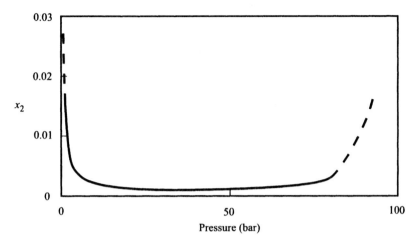

Fig. 5.35   The solubility of α-pinene in carbon dioxide at 40 °C in terms of mole fraction, x_2, calculated from the Peng–Robinson equation.

obtained by extracting a plant material could be below this amount, and although the presence of other solutes in an essential oil will improve the situation, alternative methods of trapping will need to be considered. If depressurization is used for separation, this should be carried out between 30 bar and 50 bar, preferably at 40 bar.

6 Countercurrent extraction and supercritical fluid fractionation

6.1 Introduction

For extraction from a liquid phase, countercurrent supercritical fluid extraction or supercritical fluid fractionation can be carried out. The liquid phase may be a liquid mixture, or a solution containing solutes which are solid at the extraction temperature, or a slurry of solid particles, such as microbial cells. The advantages of these processes are firstly that they are continuous and secondly that they are more efficient than a batch process, if well designed. Batch extraction is a single operation; countercurrent extraction and fractionation are processes in which there a number of *stages* or *plates*; and chromatography, described in the next chapter, has a large number of plates. Thus, in a sense countercurrent extraction and fractionation are intermediate between extraction and chromatography. The separation efficiency and costs of these processes increase in the order of extraction, countercurrent extraction/fractionation, and chromatography. It often makes economic sense to perform countercurrent extraction or fractionation as a preliminary stage to concentrate the compound of interest, before carrying out chromatography to obtain a very pure product.

Countercurrent supercritical fluid extraction and supercritical fluid fractionation involve the use of a vertical column, in which there is some internal structure or packing, and in which, because of their typical relative densities, the liquid phase descends through the column while the supercritical fluid phase ascends through it. The supercritical fluid emerging from the top of the column is a solution containing the *extract*, and the liquid leaving the bottom of the column after extraction is the *raffinate*. The difference between countercurrent supercritical fluid extraction and supercritical fluid fractionation is that the former is conducted at constant temperature, whereas in the latter a temperature gradient is imposed on the top section of the column. Not all publications use this distinction, which is convenient, however. Also in the former the liquid feed to be extracted is injected near the top of the column, whereas in the latter the feed enters at a point between the column head and base, and refluxing occurs within the column. Countercurrent supercritical fluid extraction is analogous to liquid–liquid extraction, whereas supercritical fluid fractionation has some similarities to liquid–liquid extraction with reflux and with fractional distillation.

Modelling of these processes involves the concept of a theoretical plate, that is a section of the column which results in complete equilibration of a component of interest between the two phases. The theoretical plates have a certain physical height, which can be compared with the total length of the column to give the number of theoretical plates in the column. The theory of how a plate height can be determined, which is a subject in chemical engineering, will not be discussed in this book. However, plate heights in chromatography will be discussed in the next chapter and there are some analogies. In particular, the number of plates in a column is approximately inversely proportional to the rate of flow of supercritical fluid substance through the column. Also the successful operation of a column without 'flooding' depends on the descent of the liquid phase through the column, which itself depends on a density difference between the two phases. Again the reader is referred to textbooks on chemical engineering for information on this problem. In supercritical fluid processes the densities of the liquid and supercritical fluid phase may approach each other as the pressure is raised, as described in Section 2.2.1. This effect may limit the maximum pressure and flow rates at which the process can be operated. In practice, when this is a problem, this maximum pressure will usually need to be determined by pilot-scale experiment.

6.2 Countercurrent extraction

Figure 6.1 is a schematic diagram for countercurrent extraction using a supercritical fluid. Liquid feed is pumped into the top of the extraction column, and the supercritical fluid substance is pumped from a condenser and heated to the column temperature before entering the bottom of the column. The supercritical fluid emerging from the top of the column passes through a control valve, which controls the pressure in the column, is then heated to offset the cooling effect of pressure reduction, and then passes into a separator, where the extract precipitates. The supercritical fluid substance, now at a low pressure (say 40 bar), is cooled and converted into a liquid before returning to the condenser for recycling. The raffinate and the precipitated extract emerge from the bottom of the extraction column and separator, respectively, through systems of valves and sometimes degassing chambers (not shown). Also not shown are temperature controls for the extraction vessel, separator, and condenser.

6.2.1 Modelling of countercurrent extraction

As mentioned earlier, the absolute flow rates of the feed and supercritical fluid through a column determines the efficiency of the column, which is quantified

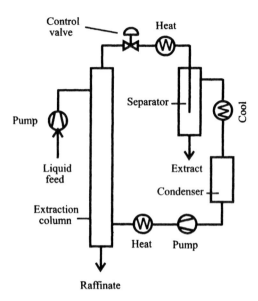

Fig. 6.1   Schematic representation of a system for countercurrent supercritical fluid extraction.

by the number of theoretical plates, N. The exact relationship is not known, although an inverse relationship between N and the flow rates is expected. The procedure that needs to be followed is therefore that the column has a given number of theoretical plates and only relative flow rates are discussed. The variables used, $A[I, J, K]$, which are elements of a matrix, are the flow rates of the M components (not counting the supercritical fluid substance) in any part of the column in moles per second, dn_i/dt, relative to that of the flow rate of the supercritical fluid substance upwards through the column in moles per second, F_x,

$$A[I, J, K] = (dn_i/dt)/F_x \qquad (6.1)$$

The matrix A is three-dimensional as follows: I is the component number, J is the plate number (numbered from the bottom), and K indicates a flow type. $K = 1$ indicates liquid entering the column, $K = 2$ liquid flowing down from the plate, and $K = 3$ fluid flowing up from the plate. The elements $A[I, J, 1]$ are thus zero except for $J = N$, as the liquid enters at the column head, where they are known. These are given in Table 6.1 which also contains the dummy matrix elements $A[I, 0, 3] = 0$ and $A[I, N + 1, 2] = 0$ as these are convenient in the calculation given below.

   The calculation will require information about the partitioning of the components between the liquid phase and the supercritical fluid phase. Any

Table 6.1   **Matrix elements, $A[I, J, K]$, for the countercurrent extraction of a liquid with two components using a column with five plates: I is the component number; J is the plate number; and K indicates a flow type, where $K=1$ is liquid entering the column, $K=2$ is liquid flowing down from the plate, and $K=3$ is fluid flowing up from the plate**

Liquid entering		Liquid flowing down		Fluid flowing up	
		$A[1, 6, 2]=0$	$A[2, 6, 2]=0$		
$A[1, 5, 1]$	$A[2, 5, 1]$	$A[1, 5, 2]$	$A[2, 5, 2]$	$A[1, 5, 3]$	$A[2, 5, 3]$
$A[1, 4, 1]=0$	$A[2, 4, 1]=0$	$A[1, 4, 2]$	$A[2, 4, 2]$	$A[1, 4, 3]$	$A[2, 4, 3]$
$A[1, 3, 1]=0$	$A[2, 3, 1]=0$	$A[1, 3, 2]$	$A[2, 3, 2]$	$A[1, 3, 3]$	$A[2, 3, 3]$
$A[1, 2, 1]=0$	$A[2, 2, 1]=0$	$A[1, 2, 2]$	$A[2, 2, 2]$	$A[1, 2, 3]$	$A[2, 2, 3]$
$A[1, 1, 1]=0$	$A[2, 1, 1]=0$	$A[1, 1, 2]$	$A[2, 1,2]$	$A[1, 1, 3]$	$A[2, 1, 3]$
				$A[1, 0, 3]=0$	$A[2, 0, 3]=0$

expression may be used for this equilibrium, but for convenience and demonstration purposes, it is here assumed that the liquid phase is an ideal mixture, that it contains only negligible amounts of the supercritical fluid substances, and that the solutes do not interact in the supercritical fluid phase. The mole fraction of each component in the supercritical fluid phase will then be equal to its solubility, in terms of mole fraction, which will be here denoted by $S[I]$, multiplied by its mole fraction in the liquid phase. At each plate and for each component therefore it will be true that

$$A[I, J, 3] = S[I] \times A[I, J, 2] / \sum A[I, J, 2] \qquad (6.2)$$

Equation (6.2) asserts that the fluid and liquid phases are in equilibrium on leaving each theoretical plate. There will be $M \times N$ equations of the type of eqn (6.2). There will also be $M \times N$ mass balance equations, which equate the quantity of each component arriving at the plate to that leaving the plate. A component arriving at the plate is included in the liquid entering, in the liquid flowing down from the plate above, and in the fluid flowing up from the plate below. A component leaving the plate is included in both the liquid and the fluid flowing from the plate, and thus

$$A[I, J, 1] + A[I, J + 1, 2] + A[I, J - 1, 3] = A[I, J, 2] + A[I, J, 3] \qquad (6.3)$$

Thus there will be a total of $2 \times M \times N$ equations, which will be exactly sufficient to determine the $2 \times M \times N$ unknown matrix elements. The output of most importance from this calculation will be the composition of the liquid flowing down from the bottom plate (the raffinate), $A[I, 1, 2]$, and that of the fluid flowing up from the Nth plate (the extract), $A[I, N, 2]$.

6.2.2   Separation of two components one of which is insoluble

A relatively simple situation is first considered in which the liquid feed consists of a component (component 1) which has very low solubility and can be considered insoluble in the supercritical phase and which is in excess, and a relatively small amount of component 2, which is soluble. A real situation which approaches this model would be the extraction of a slightly polar organic compound, such as phenol, from water using supercritical carbon dioxide. The ratio of the molar flow rate of liquid to that of the supercritical substance will not change significantly during the process and can be given a constant value F_1. Therefore $A[2, J, 2]/F_1$ will be the mole fraction in the liquid phase at plate J. For component 1 equations of the types of both eqns (6.2) and (6.3) are no longer relevant. The equilibrium for component 2 can now be expressed conveniently by a partition coefficient K_x, which is the ratio of the mole fractions of this component in the supercritical fluid phase compared with that in the liquid phase. Equation (6.2) can now be replaced for component 2 by

$$A[2, J, 3] = K_x A[2, J, 2]/F_1 \qquad (6.4)$$

Figure 6.2 shows a plot of $A[2, J, 3]$ versus $A[2, J, 2]/F_1$ for a calculation with four plates. The *equilibrium line* in this figure is a straight line of slope K_x passing through the origin. The equilibrium line is therefore a plot of the relationship given in eqn (6.4).

Substitution of eqn (6.4) into eqn (6.3) gives for the mass balance equation

$$A[2, J, 1] + A[2, J + 1, 2] + (K_x/F_1)A[2, J - 1, 2]$$
$$= A[2, J, 2] + (K_x/F_1)A[2, J, 2] \qquad (6.5)$$

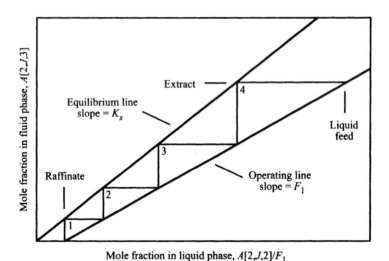

Mole fraction in liquid phase, $A[2,J,2]/F_1$

Fig. 6.2 Countercurrent extraction of a binary mixture, where component 1 is insoluble in the fluid.

After recalling that $A[2, J, 1] = 0$ for $J \neq N$, $A[2, 1, 1] = 0$, and $A[2, 0, 3] = 0$, manipulation of eqn (6.5) leads to

$$A[2, J, 3] = A[2, J + 1, 2] - A[2, 1, 2] \tag{6.6}$$

for $J = 1$ to $N - 1$ and

$$A[2, N, 3] = A[2, N, 1] - A[2, 1, 2] \tag{6.7}$$

when $J = N$. These relationships are represented on Fig. 6.2 as the *operating line*, which is a straight line of slope F_1 and an intercept on the x axis of $A[2, 1, 2]/F_1$. The four values of $A[2, J, 2]/F_1$, the liquid compositions leaving each plate, are shown as vertical lines on the staircase between the two lines. The horizontal lines on the staircase show the four values of $A[2,J,3]$, the fluid composition leaving each plate. The values of J are shown close to each pair of lines. The lines for $A[2, J, 3]$ and $A[2, J, 2]/F_1$ meet on the equilibrium line, satisfying eqn (6.4). The lines for $A[2, J, 3]$ and $A[2, J + 1, 2]/F_1$ meet on the operating line, satisfying eqn (6.6). The liquid feed composition, $A[2, 4, 1]/F_1$ ($N = 4$ in this case), is given by the x-coordinate of the point where the horizontal line for $A[2, 4, 3]$ meets the operating line. The extract composition, $A[2, 4, 3]$, is given by the highest horizontal line in the staircase, and the raffinate composition, $A[2, 1, 2]/F_1$, is given by the lowest vertical line in the staircase.

The values of the extract and raffinate compositions can be obtained by solving eqns (6.4), (6.6), and (6.7). The number of plates, N, the feed composition, $A[2, N, 1]/F_1$, the relative flow rates of feed and supercritical fluid, F_1, and the partition coefficient, K_x, are required for the calculation. For the simple system described in this section, the equations can be solved analytically and the results are as follows:

$$A[2, 1, 2]/F_1 = \frac{A[2, N, 1]/F_1}{\sum_{J=0}^{N} \left(K_x/F_1 \right)^J} \tag{6.8}$$

$$A[2, N, 3] = A[2, 1, 2] \sum_{J=1}^{N} \left(\frac{K_x}{F_1} \right)^J \tag{6.9}$$

Alternatively, the calculation may be carried out graphically using a diagram similar to Fig. 6.2. However, for a graphical calculation a different route is used. Instead of the number of plates being specified in advance and the raffinate composition obtained at the end, as in the numerical calculation described, the required raffinate composition is specified in advance and the number of plates determined. The procedure is as follows. The equilibrium line is drawn on the graph from the origin with slope K_x. The operating line is drawn from the required raffinate composition with slope F_1. The position of the operating line where the x-coordinate is equal to the feed composition is identified and a staircase drawn down to the required raffinate composition. The number of steps then gives the number of plates, which may not be integral.

This graphical method is a standard procedure in chemical engineering and forms the basis for computer programs written to model liquid–liquid extraction. More complex systems than described above can be treated and a similar procedure is used for distillation. The reader is referred to textbooks in this area, such as that by Treybal (1980). However, this procedure is not suitable and becomes too complex for supercritical fluid fractionation. This is because a temperature gradient is imposed on the column, for which the relationship with the composition inside the column is not known in advance. A unique equilibrium curve cannot therefore be drawn, and the graphical procedure and established chemical engineering related to it become too difficult. For consistency, therefore, the numerical method, described above will be used throughout this chapter and the graphs used only for explanation and illustration.

The difference between distillation, liquid–liquid extraction, and supercritical fluid fractionation can be appreciated by means of the phase rule

$$f = c - p + 2 \tag{6.10}$$

where c is the number of components, p is the number of phases, and f is the number of degrees of freedom or independent variables in the system. For the distillation of a binary mixture, $c = 2$ and $p = 2$, and so $f = 2$. If the pressure is then fixed, one degree of freedom is removed and the number of independent variables is now only one. The temperature and the compositions in the two phases can now only be changed together, that is, the temperature determines the composition in both phases. This means that (for non-azeotropic mixtures) the compositions in the liquid and vapour phases are related and a unique equilibrium curve may be drawn on a figure analogous to Fig. 6.2.

For liquid–liquid (and supercritical fluid–liquid) extraction of a binary mixture another component has been introduced, hence $c = 3$ and $p = 2$, and so $f = 3$. In this case both pressure and temperature are fixed for all compositions and the number of degrees of freedom is then reduced to one. Once again there is a fixed relationship between the compositions in both phases, and a unique equilibrium curve can be drawn as in Fig. 6.2. For supercritical fluid fractionation of a binary mixture, it is also true that $c = 3$ and $p = 2$, and $f = 3$. In this case the pressure is fixed and the number of degrees of freedom is then reduced to two. The temperature is controlled, but this is as a function of physical distance along the column and not in terms of a relationship with the compositions of the phases. A unique equilibrium curve cannot therefore be drawn.

6.2.3   General comments on countercurrent extraction

The situation discussed in the last section, where one of the components to be separated is insoluble in the fluid substance, is now used to make some generalizations about countercurrent extraction. This is done because the simple situation is easy to understand, but the generalizations made apply to more complex kinds of countercurrent separations.

Increasing the flow rate of the supercritical fluid extracting solvent compared with that of the liquid increases the running costs of the process. However, if this is done, it causes the slope of the operating line, F_1, to be lowered, reducing the number of plates and the capital cost of the plant. This is shown in Fig. 6.3, where the lowering of the operating line has reduced the number of plates from four to three for the same feed and raffinate. The general principle that increasing the flow rate of the extracting fluid relative to that of the feed increases running costs, but lowers capital costs, applies to all the processes discussed in this chapter. In practice economic analysis must be applied to determine optimum operating conditions. However, many processes are operated with the equilibrium and operating lines roughly parallel, i.e. with a value of K_x/F_1 of around unity.

It was asserted at the beginning of this chapter that a countercurrent process could be more efficient than a batch extraction, and the simple situation of the separation of two components, one of which is insoluble in the supercritical fluid will now be used to illustrate this. An example will be used in

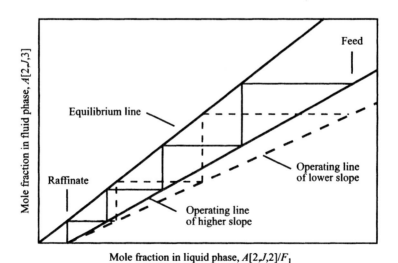

Mole fraction in liquid phase, $A[2,J,2]/F_1$

Fig. 6.3   Countercurrent extraction of a binary mixture at two different flow rates, where component 1 is insoluble in the fluid.

which there are five plates and in which $K_x/F_1 = 1$. The ratio of extraction solvent to that of liquid feed is therefore $1/K_x$. The initial mole fraction of component 2 in the feed will be $A[2, N, 1]/F_1$, and, using eqn (6.8), it can be seen that the mole fraction of component 2 in the raffinate, $A[2, 1, 2]/F_1$ will be equal to one fifth of that in the feed, i.e. $A[2, 1, 2]/F_1 = (1/5)A[2, N, 1]/F_1$.

This is now compared with a batch process in which n_1 moles of component 1, i.e. approximately n_1 of feed, is extracted with the fluid. It will be assumed that intimate contact between the liquid and the fluid is attained and thus that the extraction is only limited by solubility. The number of moles, $n_2(t)$ at any given time t will be given by

$$n_2(t) = n_2^\circ \exp(-K_x F_x t/n_1) \tag{6.11}$$

where n_2° is the initial amount and F is the flow rate of the supercritical fluid (see Section 5.5.1 for related equations, where the partition coefficient is defined in terms of volume). To reduce n_2 by a factor of five, as was achieved in the continuous extraction, $K_x Ft/n_1$ has the value of 1.6. The ratio of fluid required to that of liquid feed is therefore $1.6/K_x$, compared with $1/K_x$ in the countercurrent extraction. Apart from the 60 per cent rise in solvent requirements, the batch process requires extra time and system complexity when the cell is drained and refilled.

A half-way stage is to make the process continuous with a single stirred cell or concurrent extraction. Such processes are, however, identical to a countercurrent column with a single plate. To achieve the same result with one plate would require a K_x/F_1 value of 5, from eqn (6.8), that is to say the amount of solvent required would be five times larger.

6.2.4 Separation of a liquid into two fractions by countercurrent extraction

A more general problem is the separation of a liquid into two fractions, where all the components are soluble in the supercritical fluid to different extents. The problem is one of solving the $2 \times M \times N$ eqns (6.2) and (6.3) using numerical methods with appropriate computer software. Two limiting situations exist where the solutions of these equations are more simple. The first of these is when the liquid feed rate is very large compared with that of the extracting fluid. In this case the concentrations in the fluid extract will be approximately equal to those when the fluid is in equilibrium with a large excess of the feed, i.e.

$$A[I, N, 3] \approx S[I] \times A[I, N, 1]/\sum A[I, N, 1] \tag{6.12}$$

At the other extreme, if the liquid feed rate is very slow compared with that of the extracting fluid, such that the relative feed rates of all the components are well below their solubilities, the liquid feed will completely dissolve in the fluid and the extract concentrations will be given by the relative feed rates, i.e.

$$A[I, N, 3] = A[I, N, 1] \tag{6.13}$$

An intermediate situation can occur when some of the components have much higher solubilities than the rest and where the relative flow rates of these components are well below their solubilities. As the aim of the process is not to extract the less soluble components, a very low flow rate where eqn (6.13) is valid for them is not suitable. This situation is illustrated here by a binary mixture where one of the components in the mixture, component 2, is much more soluble and for which eqn (6.13) will apply, i.e.

$$A[2, N, 3] = A[2, N, 1] \tag{6.14}$$

Let x be the mole fraction of component 2 in the feed and f the total feed flow rate relative to that of the fluid, i.e. by definition:

$$x = \frac{A[2, N, 1]}{A[1, N, 1] + A[2, N, 1]} \tag{6.15}$$

$$f = A[1, N, 1] + A[2, N, 1] \tag{6.16}$$

Using these quantities, eqn (6.14) can be rewritten as

$$A[2, N, 3] = fx \tag{6.17}$$

An equation for component 1, valid when eqn (6.17) is valid for component 2 can be obtained as follows. For each component at plate N, eqn (6.2) can be divided by $S[I]$ to give

$$A[1, N, 3]/S[1] = A[1, N, 2]/(A[1, N, 2] + A[2, N, 2]) \tag{6.18}$$

and

$$A[2, N, 3]/S[2] = A[2, N, 2]/(A[1, N, 2] + A[2, N, 2]) \tag{6.19}$$

Adding eqns (6.18) and (6.19) together and rearranging with the use of eqn (6.17) gives

$$A[1, N, 3] = S[1](1 - fx/S[2]) \tag{6.20}$$

At high flow rates eqn (6.12) is valid for both components and using the definitions of eqns (6.15) and (6.16):

$$A[1, N, 3] = S[1](1 - x) \tag{6.21}$$

$$A[2, N, 3] = S[2]x \tag{6.22}$$

In summary, at low flow rates eqns (6.17) and (6.20) are valid and at very high flow rates eqns (6.21) and (6.22) are valid. Figure 6.4 shows plots of these equations as a function of f for $x = 0.5$, $S[1] = 0.001$, and $S[2] = 0.01$, with dotted lines for the equilibrium eqns (6.21) and (6.22) and dashed lines for the equations valid at low flow rates; eqns (6.17) and (6.20). The lines can be seen to cross and by putting the appropriate equations equal to each other, it can be shown that the crossing point is at $f = S[2]$, which has a value of 0.01 in this case. Also shown schematically by the continuous lines, are the actual values of $A[I, N, 3]$ obtained by numerical solution of eqns (6.2) and (6.3). It is found that the curves obtained are never higher than the two limiting theoretical lines for component 2, and never lower than the two limiting theoretical lines for component 1. It is also found that increasing the number of plates in a calculation brings the predicted curves nearer to the theoretical limits. As an infinite number of plates is approached, the calculated curves closely coincide with the theoretical limits, switching from one line to the other at $f = S[2]$.

The limiting equations for the extract compositions, $A[I, N, 3]$, are summarized in the first two rows of Table 6.2. As there is overall mass balance between the feed and the sum of the extract and raffinate, the raffinate compositions, $A[I, 1, 2]$, can be readily calculated for the limiting situations, and these are shown in the second and third rows of Table 6.2. From all these quantities, it is now possible to calculate purities in the fractions and these are shown in the next two rows. In this table the purity of the raffinate means the

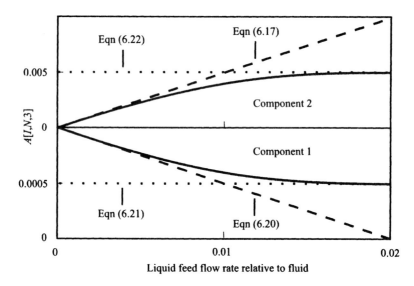

Fig. 6.4 Explanation of the separation of a binary mixture by countercurrent supercritical fluid extraction.

Table 6.2 Limiting equations for various quantities in the countercurrent extraction of a binary mixture at the two extreme values of the total feed rate relative to that of the extracting fluid, f

	$f \ll S[2]$	$S[2] \ll f$
$A[1, N, 3]$	$S[1](1 - fx/S[2])$	$S[1](1 - x)$
$A[2, N, 3]$	fx	$S[2]x$
$A[1, 1, 2]$	$f(1 - x) - S[1](1 - fx/S[2])$	$(f - S[1])(1 - x)$
$A[2, 1, 2]$	0	$(f - S[2])x$
Purity of extract	$\dfrac{fx}{fx(1 - S[1]/S[2]) + S[1]}$	$\dfrac{S[2]x}{S[1](1 - x) + S[2]x}$
Purity of raffinate	1	$\dfrac{(f - S[1])(1 - x)}{f - S[2]x - S[1](1 - x)}$
Loss of 1 in extract	$\dfrac{S[1](1 - fx/S[2])}{f(1 - x)}$	$\dfrac{S[1]}{f}$
Loss of 2 in raffinate	0	$1 - S[2]/f$

mole fraction of component 1 in the raffinate and the purity of the extract means the mole fraction of component 2 in the extract. It can be seen that although pure raffinate can be obtained at low flow rates, the purity of the extract has a maximum of

$$\frac{S[2]x}{S[1](1 - x) + S[2]x} \tag{6.23}$$

at high flow rates, which for the example of Fig. 6.4, is 0.91. Thus by using an infinite number of plates and a flow rate of $f = S[2]$, the optimum separation giving 100 mol per cent component 1 in the raffinate and 91 mol per cent of component 2 in the extract can be obtained.

Of interest also are the amounts of components lost as impurities in the fractions. These are shown in the final two rows in Table 6.2 for the limiting situations. The losses are expressed as fractions of the feed components in this table. As can be seen, at low flow rates it is possible to lose none of component 2 in the raffinate, whereas the minimum fraction of component 1 that is lost in the extract is achieved at high flow rates and is equal to $S[1]/f$. By using an infinite number of plates and a flow rate of $f = S[2]$, there would be none of component 2 lost in the raffinate and a fraction $S[1]/S[2]$ of component 1 lost in the extract, corresponding to a 10 mol per cent loss in the example of Fig. 6.4.

6.2.5   An example of the removal of a soluble component

A practical situation would be the stripping of a volatile solvent from a solute of some, but less, volatility and hence solubility. The example given here, however, is the separation of a binary mixture of 0.96 mole fraction limonene

([*R*]-4-isoprenyl-1-methyl-1-cyclohexene) with 0.04 mole fraction geranyl acetate (*trans*-3,7-dimethyl-2,6-octadienyl acetate) using supercritical carbon dioxide. This mixture is a model for lemon oil, where a small fraction consists of oxygenated compounds, valuable in perfumery, of which geranyl acetate is typical, but most of the oil consists of terpenes of low value, such as limonene. The solubility of limonene has been experimentally determined and correlated by the equation (Basile 1997):

$$\ln(x_2 p/\text{bar}) = 29.094 - 8744.4/(T/\text{K}) + 0.00945(\rho/\text{kg m}^{-3} - 600) \quad (6.24)$$

and that for geranyl acetate is given by

$$\ln(x_2 p/\text{bar}) = 66.994 - 21\,651/(T/\text{K}) + 0.02392(\rho/\text{kg m}^{-3} - 600) \quad (6.25)$$

The solubility of limonene is much higher than that of geranyl acetate under typical conditions and therefore limonene is removed in the extract. In the deterpenation of lemon oil, therefore, it is also expected that the terpenes will be removed in the extract leaving behind the oxygenates in the raffinate.

The choice of temperature and pressure conditions for extraction is not straightforward, as maximum separation occurs when the solubility ratio of the two compounds is the greatest, whereas the maximum throughput occurs when the solubility of limonene is greatest. Furthermore there are other considerations such as the thermal sensitivity of products, a minimum density difference between the two phases to allow good column operation, and process economics, which would suggest the lowest possible pressure and near-ambient temperature. For the present illustration, a pressure of 100 bar and a temperature of 327 K are chosen, which are practicable. At this pressure, the solubility of limonene relative to geranyl acetate reaches a flat maximum value of around 70 in the region of 330 K to 340 K, and the temperature chosen is at the lower limit, for thermal sensitivity reasons. Also, under these conditions there is an acceptable density difference between the liquid and supercritical fluid phases. The densities calculated from the Peng–Robinson equation of state for a binary mixture of limonene and carbon dioxide at 327 K and 100 bar are 590 kg m^{-3} for the supercritical fluid phase and 671 kg m^{-3} for the liquid phase. The small amount of geranyl acetate, which has a density of 916 kg m^{-3}, will increase this difference. It is instructive to note that the calculation also predicts that the liquid phase contains 66 mole per cent of carbon dioxide, which corresponds to 40 weight per cent under these conditions. Thus column flooding may be avoided. Under these conditions, the solubilities of limonene and geranyl acetate are 0.0194 and 0.0003512, respectively, in terms of mole fraction.

With an infinite number of plates and a relative feed flow rate equal to 0.0194, the equations of Table 6.2 allow calculation of the best result achievable. The objects of the model process are to remove as much limonene as possible from the raffinate and lose as little of the geranyl acetate in the extract. The maximum achievable therefore is 100 per cent pure raffinate and 1.8 per cent loss of geranyl acetate. Calculations were then carried out numerically using eqns (6.2) and (6.3) for five plates, i.e. the set of unknowns shown in Table 6.1. From the initial results, the important criteria of the purity of geranyl acetate in the raffinate and the loss of geranyl acetate in the extract were calculated, and these are shown in Fig. 6.5.

As the relative flow rate increases above the value of 0.0194, not all the limonene is extracted and the raffinate purity falls markedly. As the feed flow rate is reduced, however, the amount of geranyl acetate lost by extraction increases. If the relative feed flow rate is reduced to 0.015, about 15 per cent of geranyl acetate is lost in the process. In summary, too high a ratio of feed to extracting fluid does not remove all of the more soluble component, whereas too low a ratio causes too much of the less soluble component to be removed. A good compromise, as can be seen from Fig. 6.5, is a relative total feed flow rate, $A[1, 5, 1] + A[2, 5, 1]$, of 0.019, where the purity of the raffinate is almost 100 per cent, but only 2 per cent of the geranyl acetate is lost by extraction. These conditions appear to be the basis of a successful separation. They could form the basis of initial studies, although a commercial process may operate at a somewhat higher pressure to increase production rate.

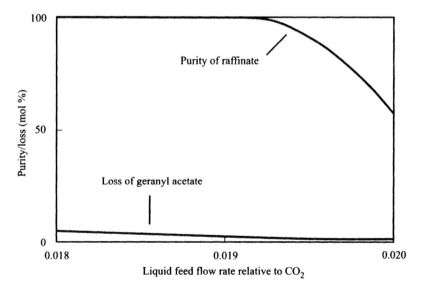

Fig. 6.5   The separation of geranyl acetate and limonene by countercurrent supercritical fluid extraction.

6.2.6 Conclusions on countercurrent extraction at constant temperature by supercritical fluids

Particular situations have been considered above, but the methods used may be generalized to more complex situations and the following conclusions drawn.

1. Countercurrent extraction is more effective in separation than single-stage extraction.

2. There is a theoretical limit to the purity of the extract, dependent on the solubilities of the components and the composition of the feed, which for a binary mixture is given by eqn (6.23).

3. There is no theoretical limit to the purity of the raffinate.

4. These theoretical limits can be approached closely by using a large number of theoretical plates in the column and by using a *total* feed flow rate relative to the fluid flow rate which is close to but usually below the solubility of the lighter component (or the average solubility of the lighter components).

5. The height of a theoretical plate is inversely dependent on the absolute flow rate. The number of plates can therefore be increased by increasing the physical size of the column or reducing the flow rates of feed and fluid.

6. The density difference between the two phases must be sufficient to prevent column flooding at a particular flow rate.

7. The principles of countercurrent extraction using a supercritical fluid are the same as for the same process using a liquid.

8. However, the larger density differences possible between the two phases may mean that a column may be operated at higher flow rates without flooding.

6.3 Supercritical fluid fractionation

6.3.1 Introduction

For substances with small solubility differences, there is a limit to the purity that can be achieved in the extract by countercurrent separation, as can be shown using the equations of Section 6.2.4 and Table 6.2. For an equimolar mixture of two components, where the solubilities differ by a factor of two, say for $x = 0.5$, $S[1] = 0.001$, and $S[2] = 0.002$, the maximum purity obtainable for component 2 in the extract is 67 mol per cent, compared with the original value of 50 mol per cent. Although this can be improved somewhat by reducing the purity of the raffinate, predictions of calculations similar to those performed in the last section are not promising. The situation occurs in liquid–liquid extraction, but in this case it is well established that the situation is improved by carrying out liquid–liquid extraction with reflux. In this latter process, solvent is evaporated from the extract and some of the residue is fed down the column to cause a 'reflux' similar to that in fractional distillation.

In the case of a supercritical fluid solvent, something of the same effect can be achieved readily by raising the temperature at constant pressure to lower the density and precipitate the extracted components, causing refluxing. This has been carried out simply by inserting a hot finger into the top of the column, but also by imposing a temperature gradient over the upper part of the column in a process defined here as supercritical fluid fractionation (Nilsson 1996). When a temperature gradient is applied in this way, the purity of the extract that can be achieved can be superior to that achieved with countercurrent extraction, if the process is properly designed. However, as a pure raffinate can be obtained with countercurrent extraction, it is not necessary to apply a temperature gradient to the lower part of the column. The lower part of the column can be described as the *stripping section* and the upper part as the *enrichment section*, which are the terms used in distillation and liquid countercurrent extraction with reflux.

Thus in supercritical fluid fractionation, the feed entry point can be located at a central point in the column and a temperature gradient imposed above the feed entry point. The temperature gradient reduces the solubility at the top of the column so that material precipitates as it rises and refluxing occurs as in a distillation. Typically the temperature gradient imposed is such that the higher temperature is at the top, causing the density to be lower there. The method works on the part of the curve where solubility is falling with temperature at constant pressure, shown in Fig. 3.2. The situation is typical for heavier solutes. In principle, the method could be used with a higher temperature at the bottom for volatile solutes. In this case the method would be very similar to distillation: it has not been described in the context of supercritical fluids, but is the situation at the head of some distillation columns in petroleum refining. When the conventional temperature gradient is used, the top temperature at the head of the column may be limited by the need to be on the falling part of the solubility curve.

Figure 6.6 shows a schematic diagram of an example of a fractionation system. The feed enters approximately centrally and the supercritical fluid enters at the bottom. One fraction emerges as a liquid at the bottom and the other as a supercritical fluid solution at the top. The lower (stripping) section is shown at a low temperature and the upper (enrichment) section is shown with a high temperature at the top. Studies of supercritical fluid fractionation have been made for the separation of olive oil, citrus oils, fish oils, and butterfat. A useful practical description of the method has been published by Nilsson (1996).

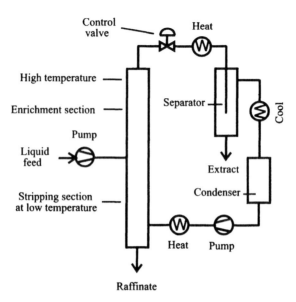

Fig. 6.6   Schematic representation of a system for supercritical fluid fractionation.

6.3.2   Modelling of supercritical fluid fractionation

As before, the calculation is carried out for M components and a given number of theoretical plates, N, and flow rates relative to those of the supercritical fluid substance are used. Now the liquid feed enters at plate L, and each plate has a particular temperature, $T[J]$, with the temperatures for the plates where $J < L$ being equal. The pressure, p, is set and the first task is to determine the density at each plate, which, in examples quoted here, is calculated from the equation of state of Span and Wagner (1996), which is inserted as a subroutine in the computer program used. Now the solubility, $S[I, J]$, will depend on the plate number, with I indicating the component and J the plate. Any solubility equation or correlation may be inserted into the calculation, for example, that described in Section 3.3, i.e.

$$\ln(S[I, J]p/p_{\text{ref}}) = a[I] + b[I]/T[J] + c[I](\rho(p, T[J]) - \rho_{\text{ref}}) \qquad (6.26)$$

which is an analogue of eqn (3.7) may be used. The simultaneous equations, modified from eqns (6.2) and (6.3) to take into account the varying solubility,

$$A[I, J, 3] = S[I, J] \times A[I, J, 2]/\sum A[I, J, 2] \qquad (6.27)$$

and

$$A[I, J, 1] + A[I, J + 1, 2] + A[I, J - 1, 3] = A[I, J, 2] + A[I, J, 3] \qquad (6.28)$$

Table 6.3   Matrix elements, $A[I, J, K]$, for the supercritical fluid fractionation of a liquid with two components using a column with five plates and the liquid feed entering at plate 3: I is the component number; J is the plate number; and K indicates a flow type with $K = 1$ is liquid entering the column, $K = 2$ is liquid flowing down from the plate, and $K = 3$ is fluid flowing up from the plate

Liquid entering		Liquid flowing down		Fluid flowing up	
		$A[1, 6, 2] = 0$	$A[2, 6, 2] = 0$		
$A[1, 5, 1] = 0$	$A[2, 5, 1] = 0$	$A[1, 5, 2]$	$A[2, 5, 2]$	$A[1, 5, 3]$	$A[2, 5, 3]$
$A[1, 4, 1] = 0$	$A[2, 4, 1] = 0$	$A[1, 4, 2]$	$A[2, 4, 2]$	$A[1, 4, 3]$	$A[2, 4, 3]$
$A[1, 3, 1]$	$A[2, 3, 1]$	$A[1, 3, 2]$	$A[2, 3, 2]$	$A[1, 3, 3]$	$A[2, 3, 3]$
$A[1, 2, 1] = 0$	$A[2, 2, 1] = 0$	$A[1, 2, 2]$	$A[2, 2, 2]$	$A[1, 2, 3]$	$A[2, 2, 3]$
$A[1, 1, 1] = 0$	$A[2, 1, 1] = 0$	$A[1, 1, 2]$	$A[2, 1, 2]$	$A[1, 1, 3]$	$A[2, 1, 3]$
				$A[1, 0, 3] = 0$	$A[2, 0, 3] = 0$

are now solved, recalling that the elements $A[I, J, 1]$ are zero except for $J = L$, and that the dummy elements $A[I, 0, 3]$ and $A[I, N + 1, 2]$ are also zero, as shown by example in Table 6.3.

Although for a particular case these equations are solved to produce a solution, there are some general principles for these calculations. These are now explained for a binary mixture in terms of the mole fractions at each plate of the more soluble component 2 in both the supercritical fluid phase, $Y[2, J]$ and the liquid phase, $X[2, J]$, which are given by

$$Y[2, J] = A[2, J, 3]/(A[1, J, 3] + A[2, J, 3]) \tag{6.29}$$

$$X[2, J] = A[2, J, 2]/(A[1, J, 2] + A[2, J, 2]) \tag{6.30}$$

Using eqn (6.27), eqn (6.29), and some algebraic manipulation, the following expression is obtained:

$$Y[2, J] = \frac{S[2, J]X[2, J]/S[1, J]}{S[2, J]X[2, J]/S[1, J] + (1 - X[2, J])} \tag{6.31}$$

As $S[2, J]/S[1, J]$ is greater than unity, $Y[2, J]$ will be greater than $X[2, J]$ and the relationship will have the form of the equilibrium curves shown in Fig. 6.7. Because there is a range of values of $S[2, J]/S[1, J]$, there will be a band of curves, and these are shown as the extreme curves with horizontal hatching between.

The operating line in the enrichment section is now considered, which, analogously to the approach used in Section 6.2.2, is the relationship between the mole fraction of component 2 in the gas phase at a particular plate to that in the liquid phase in the plate above. As before, this relationship is obtained from the mass-balance equation, eqn (6.28) in this case. At the Nth plate,

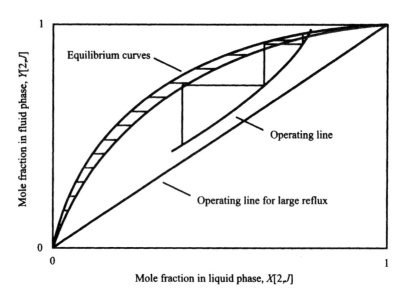

Fig. 6.7 Explanation of the separation of a binary mixture by supercritical fluid fractionation.

recalling that $A[I, N, 1] = 0$ and $A[I, N+1, 2] = 0$,

$$A[I, N-1, 3] = A[I, N, 2] + A[I, N, 3] \tag{6.32}$$

For the $(N-1)$ plate, recalling that $A[I, N-1, 1] = 0$,

$$A[I, N, 2] + A[I, N-2, 3] = A[I, N-1, 2] + A[I, N-1, 3] \tag{6.33}$$

which after substitution for $A[I, N-1, 3]$ becomes

$$A[I, N-2, 3] = A[I, N-1, 2] + A[I, N, 3] \tag{6.34}$$

Continuing this process down the column until the $(L=1)$th plate gives the general equation

$$A[I, J, 3] = A[I, J+1, 2] + A[I, N, 3] \tag{6.35}$$

for all values of J between L and N, i.e. in the enrichment section. From the definition of $Y[2, J]$ in eqn (6.29) and eqn (6.35), it is found that

$$Y[2, J] = \frac{A[2, J+1, 2] + A[2, N, 3]}{A[1, J+1, 2] + A[1, N, 3] + A[2, J+1, 2] + A[2, N, 3]} \tag{6.36}$$

For $J = L$, eqn (6.35) is, for component 2,

$$A[2, L, 3] = A[2, L+1, 2] + A[2, N, 3] \tag{6.37}$$

and a *reflux ratio* for component 2, $R[2]$, is defined by

$$R[2] = A[2, L+1, 2]/A[2, N, 3] \tag{6.38}$$

and thus is the ratio of the amount of component 2 falling down in the liquid from the plate above the entry plate to that removed in the extract. This reflux ratio for the more soluble component thus indicates the amount of refluxing occurring in the column. Conversely, a high reflux ratio means that $A[2, L+1, 2]$ is much larger than $A[2, N, 3]$. Values of $A[I, J+1, 2]$ will fall as J rises above L, but a very high reflux ratio will also mean that they are larger than $A[2, N, 3]$.

Thus, for infinitely high reflux ratios, $Y[2, J] \rightarrow X[2, J]$, according to eqn (6.36) and the definition of $X[2, J]$, given in eqn (6.30). This relationship is shown as the 'operating line for large reflux' in Fig. 6.7. For finite reflux, however, $Y[2, J] > X[2, J]$, according to eqn (6.36), and a schematic, more realistic operating line is shown above the line for infinite reflux in the figure. As can be seen, this line will cut the equilibrium curves at some point, and this will limit the maximum possible value of $Y[2, N]$ and thus the purity of component 2 in the extract. A schematic staircase is drawn on the figure to indicate the concentrations at plates in the process and, with an infinite number of plates, the point where the curves cross will be reached. Thus for high purity in the extract, the reflux ratios must be large to improve the maximum purity, and an adequate number of plates must be provided for the maximum purity to be closely approached.

Finally, if fractionation is working well, the relative flow of component 2 in the extract will be close to the solubility of this component at the highest plate. So that component 2 contaminates the raffinate to a minimum extent, the relative flow of this component in the feed must be no greater than this solubility. If it is much less, however, the reflux ratio of component 2 will be reduced. The optimum can be found by carrying out the full modelling, bearing in mind the limitations of the model and the need for eventual empirical adjustment of predicted conditions.

Equations (6.27) and (6.28) can be applied to the separation of multi-component mixtures into two fractions. The principles discussed above can be generalized to this situation, by considering each fraction to be composed of a single substance, which is either a typical component or an artificial 'average' component. The solution of the simultaneous equations is usually also carried out for a less complex model mixture. A mixture of 10 components can, for example, be treated as a model mixture of four main or typical components. This will indicate the conditions for starting laboratory or pilot-scale experiments to optimize conditions further.

6.3.3   Example calculations of supercritical fluid fractionation

The example used is that of the separation of an equimolar mixture of octadecane and hexadecan-1-ol by supercritical carbon dioxide. These are not easy substances to separate and the degree of separation that can be achieved, even with fractionation, is limited. However, this makes the problem useful for illustrating the principles of fractionation. In a real process there is usually no choice about the substances to be separated and a less than perfect separation may often be acceptable. In particular, partial separation may be acceptable as a clean-up step prior to chromatography. Much better separations can be achieved by supercritical fluid fractionation for systems where the solubilities are falling much more rapidly with temperature at constant pressure.

For the present problem of octadecane and hexadecan-1-ol separation, calculations of density and solubility were first carried out using the Peng–Robinson equation of state. The critical parameters and acentric factors used were taken from Table 1.1 for carbon dioxide and are given in Table 6.4 for the solutes. As solubilities in carbon dioxide have been published both for hexadecan-1-ol (Kramer and Thodos 1988) and octadecane (Schmitt and Reid 1988), the binary interaction parameters were obtained by fitting these solubilities, and these are also given in Table 6.4. From phase and density calculations, conditions of 150 bar, with a temperature range of 340 K to 370 K, were chosen to perform a realistic sample calculation. The choice of conditions is a compromise between high enough solubilities for a reasonable throughput and a large enough density difference between the two phases. The solubilities of the two compounds for the conditions chosen are shown in Fig. 6.8. They have a ratio in the region of 3 over the temperature range, indicating that the maximum purity of octadecane obtainable from countercurrent extraction without a temperature gradient is around 75 mol per cent.

The solubility, in terms of mole fraction, of the more soluble octadecane is 0.00259 at 370 K and, applying the arguments given in the last section, a total feed rate relative to carbon dioxide of twice this for the equimolar mixture is the maximum to be considered. For the model calculation 10 plates were

Table 6.4   Peng–Robinson parameters for octadecane and hexadecan-1-ol, used in the modelling of supercritical fluid fractionation

Parameter	Octadecane	Hexadecan-1-ol
Critical temperature (K)	749	768
Critical pressure (bar)	12.9	15.9
Acentric factor	0.800	0.829
Binary interaction parameter with CO_2	0.09	0.11

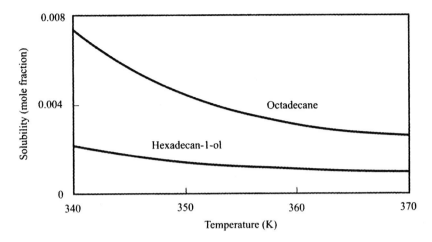

Fig. 6.8   The solubilities of octadecane and hexadecan-1-ol in carbon dioxide at 150 bar.

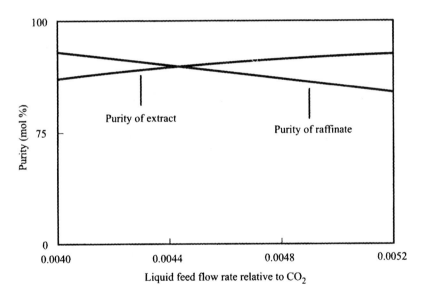

Fig. 6.9   The separation of octadecane and hexadecan-1-ol by supercritical fluid fractionation in carbon dioxide at 150 bar.

considered, with the bottom four plates at 340 K, and the temperatures of the higher plates rising by 5 K per plate to reach 370 K at plate 10. Calculations were then performed by solving eqns (6.27) and (6.28) and the purities of the two fractions obtained are shown in Fig. 6.9. As expected, the purity of the raffinate falls with increasing feed flow rate, showing similarity to what is experienced with countercurrent extraction. Conversely, the purity of the extract rises with feed flow rate, and this correlates with a rise in reflux ratio for component 2 from 0.50 to 0.62 as the relative feed flow rate rises from 0.004 to 0.0052. (The reflux ratios for component 2 are much higher than for component 1.) The choice of feed flow rate will depend on the outcome required, but for a relative total feed flow rate of 0.0044, purities in both the raffinate and extract of 90 mol per cent are obtained, and this value is used in the discussions below.

The values of $X[2, J]$ and $Y[2, J]$ obtained from the calculations are plotted in the familiar staircase form in Fig. 6.10. The coordinates of the staircase angles are then used to draw schematically an operating line and an 'equilibrium curve'; the inverted commas being used to indicate that this curve is a series of points on a number of true equilibrium curves covering the temperature range. Figure 6.10 is thus of the general form of Fig. 6.7, used to explain the principles in the last section. The operating line and equilibrium curve cross a short distance above the staircase. This indicates that the mole fraction of component 2 in the extract cannot be increased by more than 1–2 per cent by increasing the number of plates.

Finally the flows of both components in both phases, relative to the flow of carbon dioxide are shown in Fig. 6.11, with the amounts in the supercritical phase shown as open circles and the amounts in the liquid shown as bullets, and the plate numbers increasing with height. The concentration of octadecane in the supercritical phase is always above that in the liquid, especially in the enrichment section. In the stripping section, hexadecan-1-ol

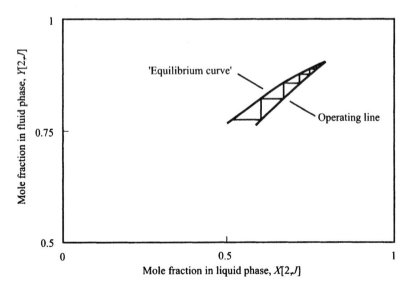

Fig. 6.10 Mole fractions of octadecane in both phases during the separation of octadecane and hexadecan-1-ol by supercritical fluid fractionation in carbon dioxide at 150 bar.

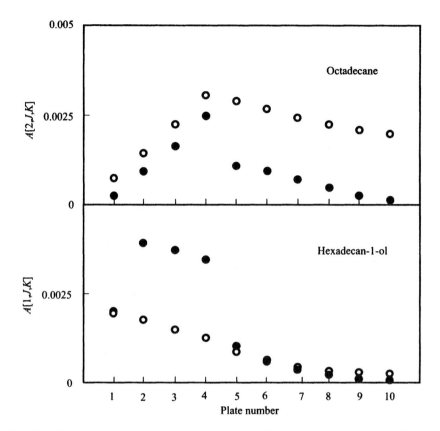

Fig. 6.11   Molar flow rates, relative to carbon dioxide, during the separation of octadecane and hexadecan-1-ol by supercritical fluid fractionation in carbon dioxide at 150 bar, with open circles representing the supercritical fluid phase and solid bullets the liquid phase.

is mainly in the liquid phase, but in the enrichment section, when the quantities are small, the concentration in the supercritical phase is marginally higher. The progress is towards separation, with the supercritical fluid phase at the top of the column being mainly octadecane and the liquid phase at the base being mainly hexadecan-1-ol.

6.3.4   Conclusions on supercritical fluid fractionation

Again more general conclusions are drawn from the discussions in this section, as follows.

1.   Supercritical fluid fractionation may improve the separation possible by countercurrent extraction, although there are still constraints to an extent depending on physical property values.

2. Better separation is obtained if the reflux ratio in the enrichment section is high, but this depends on the degree to which the solubilities of the more soluble components fall as the temperature rises at constant pressure. Conditions can be optimised to increase the solubility decrease, but this is usually limited by the need for a density difference at the high end and for a reasonable production rate at the low end.

3. For a given system under given conditions, there is a theoretical limit to the maximum purity of the extract, where an operating line crosses the equilibrium curve for the highest temperature.

4. The theoretical limit can be approached closely by enough theoretical plates in the column and by using a *total* feed flow rate relative to the fluid flow rate which is close to, but usually lower than, the solubility of the lighter component (or the average solubility of the lighter components).˙

5. The height of a theoretical plate is inversely dependent on the absolute flow rate. The number of plates can therefore be increased by increasing the physical size of the column or reducing the flow rates of feed and fluid.

6. The density difference between the two phases must be sufficient to prevent column flooding at a particular flow rate, and this will be most critical in the stripping section of the column, where the temperature is low.

7. The principles of fractionation using a supercritical fluid are different for those for fractional distillation countercurrent extraction with reflux using a liquid.

7 Supercritical fluid chromatography

7.1 Introduction and basic principles

In chromatography, compounds are dynamically partitioned between two phases, one of which is mobile, and these compounds move with the *mobile phase* to an extent depending on the fraction of time they spend in it. As different compounds will partition differently, they will travel at different speeds, and so a mixture of compounds will separate as they move. For supercritical fluid chromatography (SFC), the mobile phase is a supercritical fluid and the *stationary phase* is a packed bed of solid material, which may have specific compounds bonded on to it, or a quartz capillary wall, which is coated with a thin layer of polymer.

The fluid moves through the bed or capillary under the influence of a pressure gradient and as a result velocity gradients are set up, as described in Section 4.1. These are illustrated above for a capillary column. Because different streams of the fluid are travelling at different velocities, the compounds being transported tend to disperse, but this tendency is reduced because they move between the velocity streams as a result of radial diffusion. The dimensions of chromatographic packings and capillaries are made deliberately small to facilitate this radial diffusion. There will also be a dispersive effect due to longitudinal diffusion. As a result of the velocity, diffusive, and other effects described below, the chromatographic band for a particular substance will have a width, which will determine the *efficiency* of separation.

Diffusion in the gas phase is relatively rapid and, as a result, gas chromatography gives very narrow chromatographic bands or peaks and is very efficient. The majority of compounds are, however, not volatile enough, at temperature where they are stable, for gas chromatography to be performed on them. An alternative is to find a solvent for them and use this as a mobile phase in liquid chromatography. However, because diffusion coefficients in liquids are many orders of magnitude smaller, chromatographic peaks are much wider. Table 4.1 gives some representative values for the diffusion coefficients of naphthalene in carbon dioxide for different phases. This table shows that the diffusion coefficient in supercritical carbon dioxide is intermediate between the gas and liquid values, and, as many compounds can be dissolved in a supercritical fluid, SFC provides a useful option for efficient separation of less volatile compounds. In SFC, the density can be adjusted to give just enough solvation power and yet rapid diffusion for efficient chromatography. Many examples can be chosen to illustrate this, and here the example chosen is the separation of polystyrene of relative molar mass up to 3500. Only the lower oligomers are accessible by gas chromatography and the best form of liquid chromatography, i.e. gel permeation chromatography, will separate the first few oligomers and thereafter give only an unresolved envelope. With SFC, oligomers with up to 35 monomer units are clearly resolved, as shown in Section 7.3.1 and Fig. 7.13, below. This chromatogram is not the limit of resolution that has since been achieved.

7.1.1 Chromatography in a capillary

The easiest situation to discuss the principles of SFC initially is that of a narrow uniform capillary, typically of 50 μm internal diameter, which is coated on the inside with a thin layer of polymeric stationary phase, typically of 0.2 μm thickness. As the column has the form of a cylinder, cylindrical coordinates are appropriate and, assuming the problem has azimuthal symmetry, only the longitudinal coordinate, z, and the radial coordinate, r, are relevant. r can take on values between 0 and a, the tube radius. It is assumed initially that the pressure drop along the tube is small and that the diffusion coefficient of the solute, component 2, in the fluid is constant. The velocity gradient can be shown to be parabolic, as illustrated in the last section, and a function of r only, of the form

$$v \equiv v_z = 2v_0 \left(1 - \frac{r^2}{a^2} \right) \tag{7.1}$$

where v_0 is the average velocity, as can be shown by integrating eqn (7.1) over the tube cross-section. The concentration of solute is assumed to be small and eqn (4.30) can be used, which becomes in cylindrical coordinates, after substituting for v using eqn (7.1)

$$\frac{\partial c_2}{\partial t} = D_{12} \frac{\partial^2 c_2}{\partial z^2} + D_{12} \frac{1}{r} \frac{\partial}{\partial r} \left(r \frac{\partial c_2}{\partial r} \right) - 2v_0 \left(1 - \frac{r^2}{a^2} \right) \frac{\partial c_2}{\partial z} \tag{7.2}$$

This equation must be solved in conjunction with boundary conditions, one of which involves interactions with the stationary phase at $r = a$. These interactions will here be treated as adsorption and desorption, which is appropriate for many stationary phases in packed columns, which are the most used. It is less realistic for a capillary coating, where interaction involves solution in the stationary phase and diffusion within it. However even this interaction can be modelled as an adsorption–desorption process. Let k_1 be the first-order rate coefficient for desorption, k_2 the rate coefficient for adsorption per unit area of the surface per unit molar concentration in the fluid near the surface, and c_{st} the surface concentration of component 2. The flux of molecules near the surface, which will be given by Fick's first law (eqn (4.8)), is equal to the difference between the rates of adsorption and desorption, i.e. at $r = a$

$$-D_{12}\frac{\partial c_2}{\partial r} = k_2 c_2 - k_1 c_{st} \tag{7.3}$$

The rate of change of surface concentration will be equal to the same quantity,

$$\frac{\partial c_s}{\partial t} = k_2 c_2 - k_1 c_{st} \tag{7.4}$$

and elimination of c_{st} between eqns (7.3) and (7.4) gives

$$D_{12}\frac{\partial^2 c_2}{\partial r \partial t} + k_2 \frac{\partial c_2}{\partial t} + k_1 D_{12}\frac{\partial c_2}{\partial r} = 0 \tag{7.5}$$

This is the boundary condition sought. Other boundary conditions used relate to the distribution of component 2 at zero time. Equation (7.2) is difficult to solve because the variables are not separable, and so the moments method of Aris (1956) is used. As well as being a useful method of solution, this method gives the quantities required, as the first moment is related to the degree of retention and the second moment to the peak width. The pth moment of the distribution of the concentration of component 2, g_p, is defined by

$$g_p(r, t) = \int_{-\infty}^{\infty} c_2(r, z, t) z^p \, dz \tag{7.6}$$

To obtain equations for these moments and the appropriate boundary conditions, the differential equation, eqn (7.2) and the boundary condition, eqn (7.5), are multiplied by $z^p dz$ and integrated between $-\infty$ and $+\infty$ to give

$$\frac{\partial g_p}{\partial t} = D_{12}\frac{1}{r}\frac{\partial}{\partial r}\left(r\frac{\partial g_p}{\partial r}\right) + 2v_0\left(1 - \frac{r^2}{a^2}\right)pg_{p-1} + D_{12}p(p-1)g_{p-2} \tag{7.7}$$

and

$$D_{12}\frac{\partial^2 g_p}{\partial r \partial t} + k_2 \frac{\partial g_p}{\partial t} + k_1 D_{12}\frac{\partial g_p}{\partial r} = 0 \tag{7.8}$$

Moments integrated over the column cross-section, m_p, are also defined by

$$m_p(t) = 2\pi \int_0^a g_p(r, t) r \, dr \qquad (7.9)$$

m_0 will be the total amount of component 2 in the column, which will be expected to be constant. $m_2(t)/m_0$ will be the average distance travelled by component 2 in time t, if the integration constant is chosen so that $m_2(0)/m_0 = 0$, and will give information about chromatographic retention. $m_2(t)/m_0 - (m_1(t)/m_0)^2$ will give the mean-squared distance of component 2 from its average position at time t, i.e. the mean-squared width of the peak. Analogously, it will be assumed that $m_2(0)/m_0 = 0$, i.e. that the injected peak has no width, in setting the integration constant. Where this is not the case, the initial width will need to be added on to the width that develops.

The equation for the zeroth moment with its boundary condition, valid at $r = a$, is therefore

$$\frac{\partial g_0}{\partial t} = D_{12} \frac{1}{r} \frac{\partial}{\partial r} \left(r \frac{\partial g_0}{\partial r} \right) \qquad (7.10)$$

and

$$D_{12} \frac{\partial^2 g_0}{\partial r \partial t} + k_2 \frac{\partial g_0}{\partial t} + k_1 D_{12} \frac{\partial g_0}{\partial r} = 0 \qquad (7.11)$$

These equations are satisfied by

$$g_0 = A \qquad (7.12)$$

where A is a constant, independent of r and t. The moment averaged over the column is therefore obtained from eqn (7.9) to be

$$m_0 = \pi a^2 A \qquad (7.13)$$

and is a constant as expected.

The equation for the first moment with its boundary condition, valid at $r = a$, after substituting for g_0 is

$$\frac{\partial g_1}{\partial t} = D_{12} \frac{1}{r} \frac{\partial}{\partial r} \left(r \frac{\partial g_1}{\partial r} \right) + 2A v_0 \left(1 - \frac{r^2}{a^2} \right) \qquad (7.14)$$

and

$$D_{12} \frac{\partial^2 g_1}{\partial r \partial t} + k_2 \frac{\partial g_1}{\partial t} + k_1 D_{12} \frac{\partial g_1}{\partial r} = 0 \qquad (7.15)$$

It can be shown by substitutions that these two equations are satisfied by

$$\frac{g_1}{A} = B + \frac{v_0 a^2}{8 D_{12}} \left[\frac{r^4}{a^4} - 2 \left(\frac{1 + 2k}{1 + k} \right) \frac{r^2}{a^2} \right] + \frac{v_0}{1 + k} t \qquad (7.16)$$

where B is another constant and k is defined by

$$k = 2k_2/ak_1 \qquad (7.17)$$

It will be shown later that k is identical to the normal chromatographic capacity factor. The distance travelled by the component, averaged over the column is obtained, using eqn (7.9) and also using the condition that its initial value is zero to set the value of B, to be

$$\frac{m_1(t)}{m_0} = \frac{v_0}{1+k}t \qquad (7.18)$$

The equation for the second moment with its boundary condition, valid at $r = a$, after substituting for g_0 and g_1 is

$$\frac{\partial g_2}{\partial t} = D_{12}\frac{1}{r}\frac{\partial}{\partial r}\left(r\frac{\partial c_2}{\partial r}\right) + 2AD_{12}$$
$$+ 4v_0\left(1 - \frac{r^2}{a^2}\right)\left\{\frac{Av_0a^2}{24D_{12}}\left[3\frac{r^4}{a^4} - 6\left(\frac{1+2k}{1+k}\right)\frac{r^2}{a^2} + \frac{2+5k}{1+k}\right] + \frac{Av_0}{1+k}t\right\} \qquad (7.19)$$

and

$$D_{12}\frac{\partial^2 g_2}{\partial r\partial t} + k_2\frac{\partial g_2}{\partial t} + k_1 D_{12}\frac{\partial g_2}{\partial r} = 0 \qquad (7.20)$$

As can be shown by substitution, the solution for g_2 is the following:

$$\frac{g_2}{A} = C + \left(-\frac{k}{2+2k} + \frac{v_0^2}{2D_{12}k_1}\cdot\frac{k}{(1+k)^3} - \frac{v_0^2 a^2}{96D_{12}^2}\cdot\frac{3+16k+27k^2+20k^3}{(1+k)^3}\right)r^2$$
$$+ \frac{v_0^2}{96D_{12}^2}\cdot\frac{5+19k+17k^2}{(1+k)^2}r^4 - \frac{5v_0^2}{144D_{12}^2 a^2}\cdot\frac{1+2k}{1+k}r^6 + \frac{v_0^2}{128a^4 D_{12}^2}r^8$$
$$+ \left(\frac{2D_{12}}{1+k} + \frac{2v_0^2}{k_1}\cdot\frac{k}{(1+k)^3} + \frac{v_0^2 a^2}{24D_{12}}\cdot\frac{5+20k+21k^2}{(1+k)^3}\right)t$$
$$- \frac{v_0^2}{2D_{12}}\cdot\frac{1+2k}{(1+k)^2}r^2 t + \frac{v_0^2}{4D_{12}a^2}\cdot\frac{1}{1+k}r^4 t + \frac{v_0^2}{(1+k)^2}t^2 \qquad (7.21)$$

This is a complex expression but, after integration by application of eqn (7.9), and choice of the constant C by using $m_2(0) = 0$, the following simpler expression for $m_2(t)$ is obtained,

$$m_2(t)/m_0 - (m_1(t)/m_0)^2$$
$$= \left(\frac{2D_{12}}{1+k} + \frac{2v_0^2}{k_1}\cdot\frac{k}{(1+k)^3} + \frac{v_0^2 a^2}{24D_{12}}\cdot\frac{1+6k+11k^2}{(1+k)^3}\right)t \qquad (7.22)$$

which is discussed further in Section 7.1.2. Calculations of moments higher than 2 show that these are consistent with a gaussian peak shape at longer times.

Equation (7.18) gives the average speed of component 2 in terms of the capacity factor, k, defined by eqn (7.17). It will now be proved that k is the ratio of the total amount of component 2 on the surface of the tube to that in the fluid. The surface concentration, c_{st}, is given by eqn (7.3) to be

$$c_{st} = \frac{k_2}{k_1} c_2 + \frac{D_{12}}{k_1} \left(\frac{\partial c_2}{\partial r} \right)_{r=a} \tag{7.23}$$

The amount of component 2 on the surface in an element of tube dz is given by $2\pi a c_{st} dz$ and so the total amount on the surface, m_{st} is given by

$$m_{st} = 2\pi a \int_{-\infty}^{\infty} c_{st} \, dz \tag{7.24}$$

Multiplication of eqn (7.23) by dz and integrating between $-\infty$ and $+\infty$, using the definition of the moments in eqn (7.6) and also eqns (7.13) and (7.17) gives

$$m_{st}/m_0 = 2k_2/ak_1 = k \tag{7.25}$$

which is the relationship required. k is a product of the equilibrium constant for adsorption, k_2/k_1, and the phase ratio, which in the present case is the surface to volume ratio $2/a$.

If the time that component 2 spends in the fluid phase on average is τ, the average time spent on the surface will be $k\tau$ and the average total time in the column $(1 + k)\tau$. The peak travels at a velocity of $v_0/(1 + k)$, which is the average velocity of the fluid multiplied by a factor which is equal to the average time component 2 spends in the fluid phase divided by the total time in the column. The time spent passing through a column of length l, known as the *retention time*, t_R, will be $l(1 + k)/v_0$. The average time for the mobile phase substance to pass through the column is given the symbol t_M, which is also the time for a compound, such as the injection solvent, that is not retained in the stationary phase at all (i.e. for which $k = 0$). Therefore k may be obtained by appropriate retention measurements and is given by

$$k = \frac{t_R - t_M}{t_M}. \tag{7.26}$$

The capacity factor is also often expressed in terms of the retention volume, V_R, and the mobile phase retention volume, V_M, which are obtained by multiplying the mobile phase volume flow rate within the column by the corresponding retention times, as

$$k = \frac{V_R - V_M}{V_M} \qquad (7.27)$$

The advantage in using retention volumes is that V_M is often a constant for a particular column, although not when the fluid swells the stationary phase to an extent dependent on pressure. The disadvantage for SFC is that flow rates of the fluid substance(s) at the pump must be converted to a flow rate in the column using the densities in the column and the pump head.

7.1.2 The van Deemter equation

An alternative view of chromatographic separation, similar to that used in the last section, is to divide the column into a series of theoretical plates, within each of which complete equilibration of component 2 between the two phases occurs. Although this approach is historical in the context of analytical chromatography, it is still used in process analysis, it introduces the concept of the height of a theoretical plate, and allows chromatography and countercurrent processes to be compared. The situation in chromatography is obviously different, as one of the phases does not move and the system is not at a steady state. The problem is illustrated in Fig. 7.1, which schematically

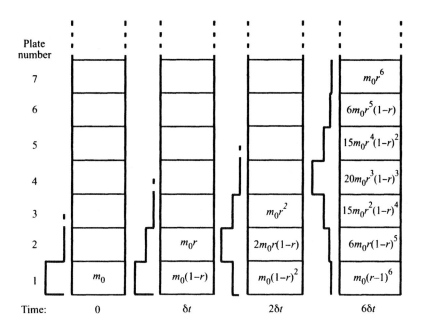

Fig. 7.1 Illustration of the plate theory of chromatography.

shows the progression of a compound along a chromatographic column, divided into a series of plates of height h. At zero time all the compound, m_0, is in plate 1, with $m_0/(1+k)$ of this in the fluid phase, with k being the capacity factor as defined in eqn (7.17) and discussed in the last section.

A time interval of δt is now considered in which the fluid moves on average a distance $v_0\delta t$, compared with a plate height of h. As only the compound in the fluid, $m_0/(1+k)$, is mobile, the amount that moves into plate number 2 will be $m_0 r$, where

$$r = v_0\delta t/h(1+k) \tag{7.28}$$

The value of δt chosen for illustrative purposes is such that r has a value of about 0.5. Thus after δt the amount of component 2 in plates 1 and 2 will be those shown in Fig. 7.1.

The fraction of the component that moves forward from one plate to the next will always be r and the fraction left behind will always be $r-1$. Therefore in the time interval from δt to $2\delta t$, $m_0 r^2$ will move into plate 3 leaving $m_0 r(1-r)$ behind in plate 2. In the same time interval $m_0 r(1-r)$ will move into plate 2 from plate 1, so that by time $2\delta t$, there will be a total of $2\,m_0 r(1-r)$ in plate 2. The amount left behind in plate 1 at the same time will be $m_0(1-r)^2$. These quantities are shown in Fig. 7.1. This process will continue, so that after $6\delta t$ the amounts in the plates will be those shown in the right-hand column diagram in Fig. 7.1. The shapes of the distributions are shown schematically and not to scale to the left of each of the column diagrams. The initial plug of material is seen to develop into the general shape of a chromatographic peak.

By extension of the arguments given above, the amount of material in the $(j+1)$th plate after a time $i\delta t$ is given by

$$m_0 \frac{i}{j!(i-j)!}r^j(1-r)^{i-j} \tag{7.29}$$

It is known that this distribution has a gaussian form at high values of i and j. The distance along the column, z, is given by jh, and the time, t, is given by $i\delta t$. Equation (7.29) can therefore be written in terms of z and t as

$$m_0 \frac{(t/\delta t)!}{(z/h)!(t/\delta t - z/h)!}r^{z/h}(1-r)^{t/\delta t - z/h} \tag{7.30}$$

It can be shown that the first moment of this distribution in terms of distance for a given time is given by

$$\frac{m_1(t)}{m_0} = \frac{v_0}{1+k}t \tag{7.31}$$

which is identical to that obtained by the previous approach and given by eqn (7.18). The second moment can be shown to be given by

$$m_2(t)/m_0 - (m_1(t)/m_0)^2 = \frac{v_0 h}{1 + k} t \qquad (7.32)$$

which is identical to eqn (7.22) provided that

$$h = \frac{2D_{12}}{v_0} + \frac{v_0 a^2}{24 D_{12}} \cdot \frac{1 + 6k + 11k^2}{(1 + k)^2} + \frac{2v_0}{k_1} \cdot \frac{k}{(1 + k)^2} \qquad (7.33)$$

An equation similar to this was first published by Golay (1958). It is one of a number of versions of the general form

$$h = A + B/v_0 + C_m v_0 + C_{st} v_0 \qquad (7.34)$$

of which the first version of a general form, in the context of chromatography, was published by van Deemter *et al.* (1956). Equation (7.34) is usually known, therefore, as the van Deemter equation. For a capillary column, $A = 0$, the term in B arises from longitudinal diffusion in the mobile phase, the term in C_m arises from the velocity gradient in the mobile phase modified by radial diffusion, and the term in C_{st} arises from slow interchange with the stationary phase. In the approach used earlier and in eqn (7.33), the slow interchange is represented by a finite rate coefficient for desorption, k_1. If this is modelled as solution into a stationary-phase layer of thickness d_{st}, diffusion within the layer quantified by a diffusion coefficient, D_{st}, followed by solution into the mobile phase, then this term becomes modified to

$$C_{st} v_0 = \frac{2 d_{st}^2 v_0}{3 D_{st}} \cdot \frac{k}{(1 + k)^2} \qquad (7.35)$$

For a packed column, the van Deemter equation has similarity with that for a capillary column, since the passages through a packed column can be regarded as capillaries of variable geometry. However, most of the terms are changed. A now becomes a finite unknown constant and this is thought to be due to the fact that the passages through the packed bed are of differing lengths, thus adding to peak broadening. These differing path lengths will however reduce longitudinal diffusion, because longitudinal concentration gradients are reduced, again by an unknown factor, γ. Because of this the B term becomes modified to

$$B/v_0 = 2 \gamma D_{12}/v_0 \qquad (7.36)$$

Because the physical causes behind the A and B/v_0 terms have something in common, they are believed to interact and as a results the A term is sometimes given as $A v_0^{1/3}$, in which case eqn (7.34) is known as the Knox equation (Knox and McLaren 1964).

For a packed capillary, the $C_m v_0$ term is given as a function of the particle size instead of the capillary radius and becomes

$$C_m v_0 = \frac{v_0 d_p^2}{24 D_{12}} \cdot \frac{1 + 6k + 11k^2}{(1+k)^2} \qquad (7.37)$$

The $C_{st} v_0$ terms can have either of the two versions above. For SFC with packed columns the Horvath–Lin equation, given in simplified form by Lee and Markides (1990), is often used, i.e.

$$h = 1.5 d_p + \frac{1.4 D_{12}}{v_0} + \frac{2(\beta + k + \beta k)^2 d_p^2 v_0}{15(1+\beta)^2 (1+k)^2 D_{12}} \qquad (7.38)$$

where β is the ratio of particle pore volume to the volume in the column not occupied by the particles. Thus this equation takes into account diffusion within the pores of the stationary phase particles, which is assumed to be the principle dispersive effect of the stationary phase. It is of the general van Deemter form with $C_{st} v_0$ and $C_m v_0$ combined into a single $C v_0$ term. Thus various expressions are used in the van Deemter equation in different circumstances. However, there is common behaviour with respect to their dependence on D_{12}. A terms are independent of D_{12}, B terms are dependent on D_{12} to the first power, and C_m terms depend on D_{12} to an inverse first power. If the C_{st} term is small, as is often the case, then the whole C term is inversely dependent on D_{12} to the first power.

Figure 7.2 shows how the terms in the van Deemter equation add up to give the form of the plate height as a function of average mobile phase velocity, v_0. In this figure again there is one combined $C v_0$ term. It can be seen that the plate height has a shallow minimum, indicating that there is an optimum average velocity, v_{opt}. At this point longitudinal diffusion and dispersion by the velocity gradient countered by radial diffusion reach a compromise. Differentiation of the van Deemter equation, eqn (7.34), setting the result equal to zero, and using the arguments on the general form of B and C given above, gives

$$v_{opt} = \sqrt{B/C} \propto D_{12} \qquad (7.39)$$

Diffusion coefficients in supercritical fluids are one order of magnitude above those in liquids, as described in Section 4.1, partly because the density is lower and partly because typically supercritical fluid substances are composed of light mobile molecules. Consequently v_{opt} is larger in supercritical fluids and therefore analysis and process times can be shorter.

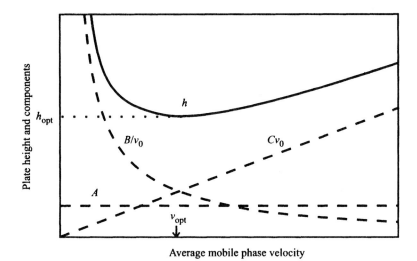

Fig. 7.2 Plot of the van Deemter equation for plate height and its components.

The minimum plate height, h_{min}, can be found by substituting v_{opt} into the van Deemter equation to be

$$h_{min} = A + 2\sqrt{BC} \qquad (7.40)$$

and is thus approximately independent of D_{12}. Thus there is no theoretical improvement in plate height in moving from liquid chromatography to SFC. However, because the $C_{st}v_0$ term is smaller in SFC, the increase in plate height on moving to a higher mobile phase velocity than v_{opt} is less than that obtained with liquid chromatography. SFC is therefore often carried out with mobile phase velocities higher than v_{opt}. These trends are summarized in Fig 7.3, which shows two schematic van Deemter curves for liquid chromatography and SFC. The parameters for both curves are the same except that the parameter B used for SFC is a factor of 10 higher, and the parameter C 10 times less for SFC. The figure shows clearly that SFC can be carried out more rapidly with less loss of efficiency.

7.1.3 Efficiency, selectivity, and resolution

Both approaches used above indicate that the distribution in a chromatographic peak should be gaussian. This will be true as a function of distance along the column and it will be approximately true as a function of time for the concentration of the component eluting from a column of given length, l. With this information, h can be obtained from the width of the peak, which is conveniently measured at 'half-height', i.e. half way between the top of the

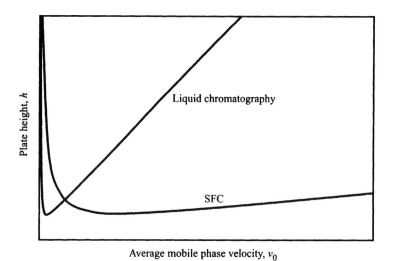

Fig. 7.3   Schematic comparison of plate height versus mobile phase velocity for liquid chromatography and SFC, where the diffusion coefficient in the mobile phase has been assumed to be a factor of 10 higher.

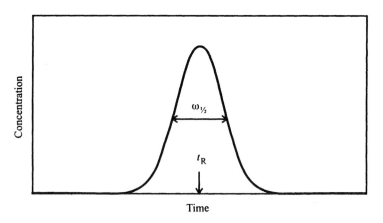

Fig. 7.4   A chromatographic peak, showing the peak parameters.

peak and the baseline, as shown in Fig. 7.4. This width, $w_{1/2}$ is expressed in terms of time, but in terms of distance is, $w_{1/2}v_0/(1 + k)$, as obtained by multiplying by the average speed of the component. The second moment of a gaussian distribution, using the notation of the last two sections is related to the width at half-height by

$$m_2(t)/m_0 - (m_1(t)/m_0)^2 = 5.545[w_{1/2}v_0/(1+k)]^2 \qquad (7.41)$$

which, after recalling that the time spent passing through a column of length l, t_R, is $l(1+k)/v_0$, can be rewritten, for $t = t_R$, as

$$m_2(t)m_0 - (m_1(t)/m_0)^2 = 5.545(w_{1/2}l/t_R)^2 \qquad (7.42)$$

Again using the definition of t_R, eqn (7.32) can be written, for $t = t_R$, as

$$m_2(t)/m_0 - (m_1(t)/m_0)^2 = hl \qquad (7.43)$$

and thus, by eliminating the second moment expression from eqns (7.41) and (7.43),

$$h/l = 5.545(w_{1/2}/t_R)^2 \qquad (7.44)$$

The quantity h/l is also equal to $1/N$, where N is the number of plates in the column. Equation (7.44) thus allows the plate height or the number of plates to be obtained from experimental chromatographic data.

Where there are two components, 2 and 3, to be separated, each will have its own capacity factor, k_2 and k_3, respectively. A *selectivity factor*, α, is defined by $\alpha = k_3/k_2$, choosing the naming of the components such that $\alpha > 1$. The resolution, R, is defined for the present purposes by

$$R = \frac{N^{1/2}}{4}\left(\frac{\alpha - 1}{\alpha}\right)\left(\frac{k_3}{1 + k_3}\right) \qquad (7.45)$$

in which it is assumed that the substances are sufficiently similar that N is essentially the same for both and k_2 and k_3 are close in value. For analytical purposes a value of R greater than 2 is often considered acceptable and this is approximately the situation illustrated in Fig. 7.5. Separation relies on either or both high efficiency , i.e. high N, or good selectivity, i.e. high α.

If resolution is good for low concentrations, it may be advantageous, for preparative or process chromatography, to *overload* the column which reduces resolution but increases production rate. When the column is overloaded, partitioning between the mobile and stationary phases is no longer proportional and expressed by the capacity factor, k. Different types of adsorption behaviour can occur, leading to different shapes of the chromatographic peak. For the present purposes, it is sufficient to note that the chromatographic peak becomes broader and its shape becomes less gaussian. If a length of column δl is required to adsorb all the component in a peak, the plate height cannot be less than δl.

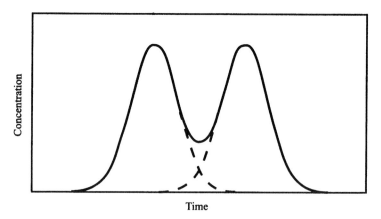

Fig. 7.5 Resolution of two chromatographic peaks.

7.1.4 Preparative and process-scale chromatography

Because SFC is more rapid than liquid chromatography it may have economic advantages for preparation and production, in spite of the somewhat higher cost of capital equipment. In addition to the narrower peaks obtained because of faster diffusion, lower viscosities mean that fast flow rates can be obtained with lower pressure drops across the column. There is an additional advantage for supercritical fluids in simulated moving bed chromatography, described below in Section 7.4. SFC is usually applied to materials of high value that are difficult to separate, such as chiral mixtures of pharmaceutical interest. The parameters that can be varied in preparative SFC, those of column type, pressure, temperature, and modifier concentration, are best chosen by observations of separation on the chromatograms obtained, often conveniently, on an analytical scale. Flow rate and loading can then be more systematically studied, as is now described.

The aim of preparative and process chromatography is to optimize production rate for a given purity. A mixture of two compounds is considered in which there are equal masses of both. If their two peaks are well separated, a plot of purity versus the fraction of total material collected will appear as curve A in Fig. 7.6. The first half collected will be one pure compound, which will be progressively contaminated by the second compound as it is added in to the material collected. If the separation is worse, curve B will be obtained, with the second compound emerging before all the first compound has been collected. If separation is poor, curve C is obtained, with some of the second compound emerging initially. For a given separation under a given set of other conditions, worse separation is obtained if either the flow rate or loading is increased. At the same time, however, the production rate is increased. Thus, if

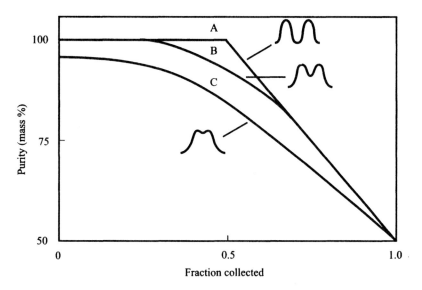

Fig. 7.6 Purity versus fraction collected for different degrees of resolution.

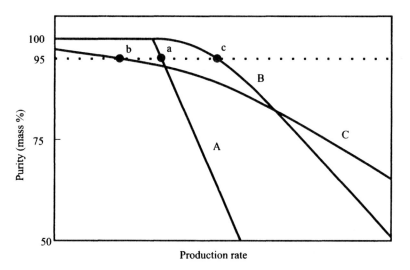

Fig. 7.7 Production rate versus purity in preparative chromatography.

purity is plotted against production rate, the curves A, B, and C of Fig. 7.6 will become spread out as in Fig. 7.7. This is the form in which data are commonly presented, as shown below. The horizontal dashed line in Fig. 7.7 represents a purity of 95 mass per cent and it cuts the curves A, B, and C at the points a, b, and c. These points are also shown on Fig. 7.8 as a schematic plot of production rate versus loading or flow rate for a purity of 95 mass per cent. Figure 7.8 thus shows how an optimum loading or flow rate arises for a given required purity.

These curves are now shown for the separation of milbemycin α_2 from a toluene extract of microbial cells (Bartle *et al.* 1995). This extract contained only ~2.5 mass per cent of milbemycin α_2 (eluted first), other milbemycins, and other compounds. The maximum production of milbemycin α_2 of 90 per cent purity obtained in a single run was 10 mg using a column 250 mm in length with a 20 mm internal diameter packed with cyanopropyl bonded 5 μm silica particles, a temperature of 320 K, a pressure of 120 bar, and a mobile phase of 5 per cent methanol in CO_2 measured as volume per cent at the pumps. Figures 7.9 (a) and (b) show examples of chromatograms obtained from the detector of the preparative SFC system at 0.1-g and 0.5-g loadings of crude extract, respectively, using a 20 ml per minute flow rate of CO_2. This figure also shows the fractions collected, during the run. These fractions were analysed by HPLC, and from their purities and times of collection, curves of purity versus production rate were obtained, by progressively adding in these fractions. These are shown in Fig. 7.10, for a range of loadings and are seen to be similar in form to the curves of Fig. 7.7. A similar series of curves could be obtained by increasing the flow rate.

These curves can then be further analysed, to obtain optimum loading for a given purity of milbemycin α_2. Below 100 per cent purity, the curves of Fig. 7.10 are fitted to a quadratic and the equations obtained are used to calculate values of the production rate at a given purity for the various loadings. Curves can then be drawn of production rate versus loading for a given

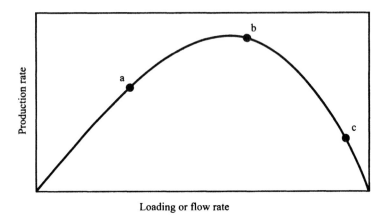

Fig. 7.8 Schematic illustration of optimum loading or flow rate during preparative chromatography.

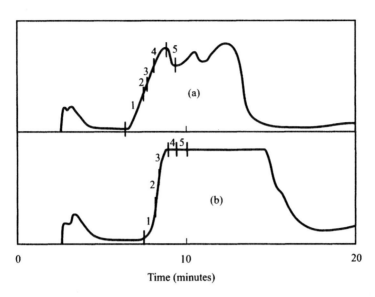

Fig. 7.9   SFC chromatograms of an extract containing milbemycin α_2 obtained during preparative chromatography at loadings of (a) 0.1 g and (b) 0.5 g.

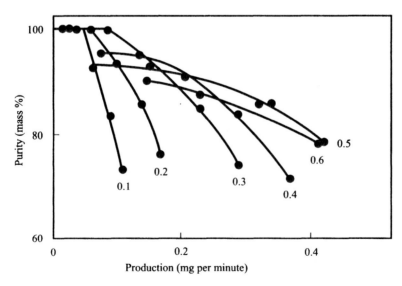

Fig. 7.10   Experimental measurements of purity versus the production rate for the preparative chromatography of milbemycin α_2 at various loadings, given in grammes on the figure.

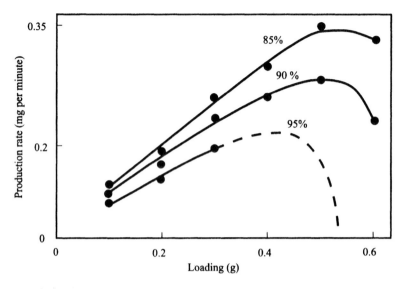

Fig. 7.11 Experimental measurements of production rate versus loading for the preparative chromatography of milbemycin α_2 for various required purities.

purity and this is shown in Fig. 7.11, with a schematic extension of the 95 per cent purity curve. As can be seen, there is an optimum loading, for a certain required purity. Similar curves can be obtained for the effect of flow rate. Theoretically, a two-dimensional optimization for both loading and flow rate could be carried out, although errors in the experimental data make this difficult and an iterative procedure appears to be better.

7.2   Thermodynamic basis for the capacity factor

At equilibrium at constant pressure, p_{chr}, and temperature, T, equating the chemical potentials of the solute in the mobile and stationary phases gives

$$\mu_{\text{m}}^{\ominus} + \int_{0}^{p_{\text{chr}}} V_{2,\text{m}}\,\mathrm{d}p + RT\ln a_{\text{m}} = \mu_{\text{st}}^{\ominus} + \int_{0}^{p_{\text{chr}}} V_{2,\text{st}}\,\mathrm{d}p + RT\ln a_{\text{st}} \qquad (7.46)$$

where $\mu_{\text{m}}^{\ominus}$ and $\mu_{\text{st}}^{\ominus}$ are the standard chemical potentials of the solute in the mobile and stationary phases, respectively, and refer to infinite dilution and standard pressure. The solute will be component 2 as before, but because all the thermodynamic quantities used in this section are for component 2, the subscript 2 will be omitted from the symbols except from partial molar volumes. In all cases, but importantly in the case of a supercritical fluid, the state at the standard pressure of one atmosphere is a hypothetical and ideal one, where the activity or fugacity coefficient, defined as being unity in the limit of zero pressure, is also unity in the standard state. $V_{2,m}$ and $V_{2,st}$ are the partial molar volumes at infinite dilution of the solute, and a_m and a_{st} are the activities of the solute in the corresponding phases.

Rearrangement of eqn (7.46) gives

$$\mu_m^{\ominus} - \mu_{st}^{\ominus} + \int_0^{p_{chr}} (V_{2,m} - V_{2,st})\, dp = RT\ln(a_{st}/a_m) \qquad (7.47)$$

If we assume that chromatography is carried out under conditions which approximate to infinite dilution, the activities can be replaced in eqn (7.47) by the concentrations divided by standard concentrations, $c_m/c_m^{\ominus}$ and $c_{st}/c_{st}^{\ominus}$. The standard concentrations are shown as different, in general, for the two phases, as will necessarily be the case for adsorption chromatography, where c_{st} and $c_{st}^{\ominus}$ will be surface concentrations. The standard concentration in the mobile phase will be taken as 1 mol per unit volume, i.e. $c_m^{\ominus} = 1/V$, where V is the molar volume in the supercritical mobile phase, which can be taken as the molar volume of the supercritical fluid substance in the limit of infinite dilution. Analytical chromatography is normally carried out at very low dilution and, if this is not the case, it is observable as a distortion of the chromatographic peak shape and a variation of the degree of retention with the amount of solute injected. The ratio of these concentrations is related to the capacity factor, k, by

$$k = \beta(c_{st}/c_m) \qquad (7.48)$$

where β is the stationary to mobile phase ratio, introduced in Section 7.1.1, where it was shown to be equal to the surface to volume ratio for adsorption chromatography as well as for chromatography where a bonded stationary phase is used. In the case of partition chromatography, when the standard concentrations in the two phases are identical, it will be the ratio of the volumes of the two phases. Substitution of eqn (7.48) into eqn (7.47), with the activities replaced by concentration ratios, gives the following expression:

$$\mu_m^{\ominus} - \mu_{st}^{\ominus} + \int_0^{p_{chr}} (V_{2,m} - V_{2,st})\, dp = RT\ln(k/\beta V c_{st}^{\ominus}) \qquad (7.49)$$

The partial molar volume of the solute in the stationary phase, $V_{2,st}$, will be close to the molar volume of the solute at all pressures and will now be assumed to be equal to it and constant. Its partial molar volume in the mobile phase, $V_{2,m}$, will vary widely, especially in the critical region, as described in Section 2.10. Its integral with pressure is given by eqn (2.12), which in the notation of the present problem and after substitution for the fugacity, is given by

$$\int_0^{p_{chr}} V_{2,m} \, dp = RT \ln(\phi_{2,m} \, p_{chr}/p^{\ominus}) \qquad (7.50)$$

Using these arguments, including eqn (7.50), eqn (7.49) can be rearranged to

$$\ln k = \ln(\phi_{2,m} V c_{st}^{\ominus} \beta p_{chr}/p^{\ominus}) - (p_{chr} V_{2,st} + \mu_{st}^{\ominus} - \mu_m^{\ominus})/RT \qquad (7.51)$$

The terms $(\mu_{st}^{\ominus} - \mu_m^{\ominus})$ are equal to the standard Gibbs function change for the transfer of the solute from the vapour to the stationary phase, $\Delta G_{chr}^{\ominus}$, and so the equation for k can be written finally as

$$\ln k = \ln(\phi_{2,m} V c_{st}^{\ominus} \beta p_{chr}/p^{\ominus}) - (p_{chr} V_{2,st} + \Delta G_{chr}^{\ominus})/RT \qquad (7.52)$$

7.2.1 Relationship between capacity factor and solubility

The solubility of component 2, expressed as concentration, $S = x_2/V$, at the same pressure and temperature as chromatography is given by eqn (3.10), which, using the notation of the last section, and taking $V_{2,st}$ to equal the molar volume of component 2, becomes

$$\ln S = -\ln(\phi_{2,m} p_{chr} V/p_v) + p_{chr} V_{2,st}/RT \qquad (7.53)$$

where p_v is the vapour pressure of component 2 at temperature T. Addition of eqns (7.52) and (7.53), making the assumption that $\phi_{2,m}$ is the value for infinite dilution, shows that

$$\ln k = -\ln S + \ln(c_{st}^{\ominus} \beta p_v/p^{\ominus}) - \Delta G_{chr}^{\ominus}/RT \qquad (7.54)$$

For a particular temperature, column and component, it can be assumed that the second and third terms on the right-hand side of eqn (7.54) are constant, in which case

$$k = C/S \qquad (7.55)$$

where C is a constant. Thus the capacity factor is inversely proportional to solubility and previous discussions of solubility behaviour can be used to explain retention behaviour.

The assumptions made in this section are not always valid. A polymer coating inside a capillary may swell increasingly as the pressure of the fluid mobile phase is raised, causing β and $\Delta G_{chr}^{\ominus}$ to change with pressure. Also solute molecules may associate in a saturated solution, so that $\phi_{2,m}$ in eqn (7.53) will not be at its infinite dilution value, even though the solution is dilute. However, in many cases the relationship is a good one. Figure 7.12 shows some results for naphthalene in carbon dioxide, with the inverse of

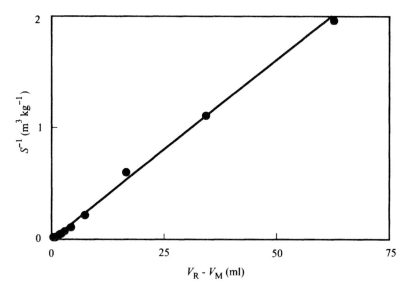

Fig. 7.12   Graph to illustrate the inverse relationship between solubility and retention for naphthalene in carbon dioxide.

the solubility (Tsekhanskaya *et al.* 1964) plotted against retention volumes (Bartle *et al.* 1990*b*). A packed column was used for the retention results which contained silica particles bonded with octadecylsilyl groups. An approximately linear relationship is obtained over a wide range, with the highest results a factor of ~500 times the lowest results. Even when conformation to the inverse linear relationship is worse, solubility is a good basis for qualitative discussion of retention. The relationship also provides a method of rapidly producing a wide range of solubility data (Cowey *et al.* 1995).

7.3   The effect of conditions on SFC

Probably the most important choice in carrying out SFC is the stationary phase, which can be less systematic than the choice of other conditions. A range of columns are available, which in the case of packed columns were mainly developed for liquid chromatography. The action of the stationary phase depends on interactions with the particular substances being separated and column types include non-polar, polar, chiral, and liquid crystal, the latter being sensitive to molecular shape. There is a great deal of analytical SFC literature, and textbooks are available which summarize the work, such as that by Lee and Markides (1990). In liquid chromatography, the term *normal phase* refers to the use of a polar stationary phase, such as silica, and a non-polar mobile phase, such as a hydrocarbon. *Reverse phase* refers to the use of a non-polar stationary phase, such as silica bonded with octadecylsilyl (ODS) groups, and a polar mobile phase, such as a water–methanol mixture. In SFC these terms are less applicable and carbon dioxide, which would be classified as a non-polar mobile phase, is often used with the non-polar ODS stationary phase.

7.3.1   Effect of pressure

The effect of increasing pressure at constant temperature is to increase solubility and therefore to decrease retention. Separation for two similar substances will depend both on small differences in the interaction with the stationary phase and also on small differences in solubility. The differences in solubility will be greatest on the rising part of the solubility curve, shown in Fig. 3.1. Pressure is therefore chosen to be low initially so that the solubility differences for the least retained and most soluble components in the mixture are high. However, this means that the heavier, least soluble components will only emerge after an unacceptably long time, and therefore pressure is often increased during chromatography. This process is known as pressure or density programming and is the equivalent of temperature programming used in gas chromatography.

The effect is most easily illustrated for the SFC of a polymer (Fjeldsted *et al.* 1983). Their approach was used earlier in Section 5.5.6 for the extraction of a polymer, and the dependence of solubility on polymer chain length was given in eqn (5.49). If the inverse of this relationship is used for the capacity factors, then

$$k_i = B_1 \exp[B_2 - B_3\rho(t)]i \qquad (7.56)$$

where k_i is the capacity factor for the *i*th oligomer. The aim is for the oligomers to emerge at equal intervals, in which case good separation can be obtained consistent with speed. The arguments are complex, since the chromatography of an individual oligomer will take place over a range of density, and so qualitative arguments are given here.

For constant density, the k_i can be seen to rise exponentially with chain length, so that oligomers will emerge more and more slowly. For a linear density gradient with time, as with polymer fractionation by extraction, the oligomers will emerge more and more rapidly and eventually a density may be reached when all oligomers are equally soluble. The use of what is described as an *asymptotic density programme* is to obtain the desired result of equally spaced peaks. This is illustrated in Fig. 7.13 for the chromatography of polystyrene in pentane using a 10 m open tubular column with an internal diameter of 100 μm coated with poly(methyl-*co*-methylsiloxane) at 210 °C.

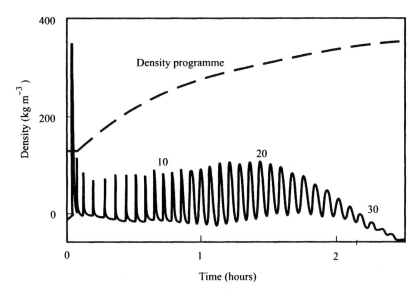

Fig. 7.13 Asymptotic density programming for the chromatography of polystyrene in pentane (Fjeldsted *et al.* 1983), redrawn by permission of the authors.

The density is seen to rise initially more rapidly, but the rate of rise decreases with time, and the peaks emerge at regular intervals. Because of the relationship between pressure and density under these conditions, a linear pressure programme produces a density programme of approximately the same shape and gives a similar result, except at low pressures (Fjeldsted *et al.* 1983).

The flow of fluid produces a pressure drop along a column, although this is smaller for SFC than for liquid chromatography, because of the lower viscosity. Chromatograms and data produced are usually referred to the mean pressure in the column. In the critical region, this error can be very large. Figure 7.14 shows the per cent deviation of the observed capacity factor from the capacity factor at the mean pressure for phenanthrene in carbon dioxide using an ODS column at 308 K, 4 K above the critical temperature of CO_2. Artificial but realistic data were used to perform the calculations, which were made for 5 bar and 30 bar pressure drops across the column. However, at densities removed from the critical density, the effect is much smaller. At higher temperatures the effect is also reduced and moves to higher pressures, corresponding to the critical density.

7.3.2 Effect of temperature

Because solubility, as a function of temperature at constant pressure, goes through a minimum, the capacity factor has a maximum at a certain

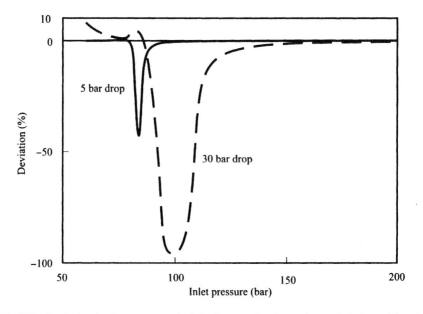

Fig. 7.14   Graph showing the percentage deviation between the observed capacity factor and that at the average pressures, obtained using artificial but realistic data for phenanthrene in carbon dioxide.

temperature. Thus increasing the temperature may increase or reduce retention. This is shown in Fig. 7.15 for some polyaromatic hydrocarbons chromatographed at 130 bar using CO_2 in an ODS column (Bartle *et al.* 1988). These data can be analysed by rearranging eqn (7.52) to give, after substituting $\Delta H_{chr}^{\ominus} - T\Delta S_{chr}^{\ominus}$ for $\Delta G_{chr}^{\ominus}$,

$$\ln k - \ln(\phi_{2,m}) - \ln(Vc_{st}^{\ominus}\beta) = (p_{chr}/p^{\ominus}) + \Delta S_{chr}^{\ominus}/R$$
$$- (p_{chr}V_{2,st} + \Delta H_{chr}^{\ominus})/RT \quad (7.57)$$

The quantity $c_{st}^{\ominus}\beta$ is included in the term $\ln(Vc_{st}^{\ominus}\beta)$ to make the argument dimensionless and it is taken here to be 1 mol m^{-3}; its value is not important as only the slopes of the graphs discussed below are important and it does not affect these. Thus if it is assumed, as is often the case in chemical problems that $\Delta H_{chr}^{\ominus}$ and $\Delta S_{chr}^{\ominus}$ do not vary much with temperature, a plot of $\ln k - \ln(\phi_{2,m}) - \ln(Vc_{st}^{\ominus}\beta)$ against $1/T$, should give a straight line with a slope of $-(p_{chr}V_{2,st} + \Delta H_{chr}^{\ominus})/R$.

Figure 7.16 shows plots versus $10^3/T$ of the various elements of this quantity: the open circles being the experimental values of $\ln k$ for fluorene, previously shown in Fig. 7.15; the continuous curve being values of $-\ln(\phi_{2,m})$, calculated from the Peng–Robinson equation of state; and the dashed line

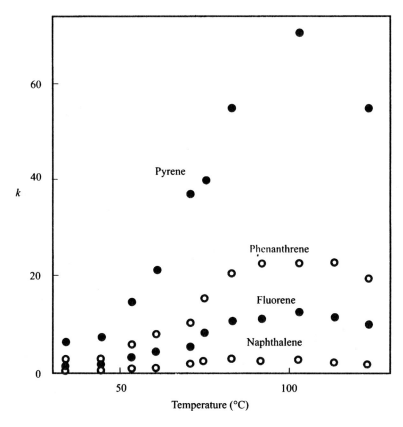

Fig. 7.15 The relationship between capacity factor, k, and temperature at a constant pressure of 130 bar for some polyaromatic hydrocarbons chromatographed in carbon dioxide.

being values of $-\ln(Vc_{st}^{\ominus}\beta)$ obtained for pure carbon dioxide from the Span and Wagner equation of state (1996). The sum of these quantities, $\ln k - \ln(\phi_{2,m}) - \ln(Vc_{st}^{\ominus}\beta)$, is shown as the filled circles and can be seen to be approximately a straight line. The value of $-(p_{chr}V_{2,st} + \Delta H_{chr}^{\ominus})$, obtained from the line is given in Table 7.1, along with values obtained from the results for the other polyaromatic hydrocarbons shown in Fig. 7.15. These are seen to be close to the corresponding enthalpy of vaporization.

7.3.3 Effect of modifiers

The use of modifiers allows more polar molecules to be chromatographed using carbon dioxide. Even salts, particularly if both ions are organic molecules, can sometimes be eluted in a CO_2–modifier mixture. Modifiers were discussed in Section 2.2 and the one most commonly used is methanol.

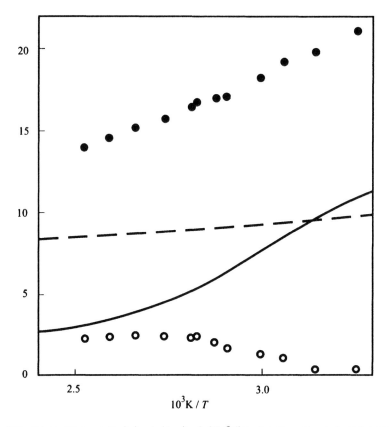

Fig. 7.16 Values of the quantity $\ln k - \ln(\phi_{2,m}) - \ln(Vc_{st}^{\ominus}\beta)$ for the chromatography of fluorene in carbon dioxide (filled circles), with its components $\ln k$ (open circles), $-\ln \phi_{2,m}$ (continuous line), and $-\ln(Vc_{st}^{\ominus}\beta)$ (dashed line).

Table 7.1 Values of the quantity $-(p_{chr}V_{2,st} + \Delta H_{chr}^{\ominus})$ for some polyaromatic hydrocarbons, compared with their enthalpies of vaporization, $\Delta H_v^{\ominus}$

	$-(p_{chr}V_{2,st} + \Delta H_{chr}^{\ominus})$ (kJ mol^{-1})	$\Delta H_v^{\ominus}$ (kJ mol^{-1})
Naphthalene	70 ± 3	70
Fluorene	82 ± 3	83
Phenanthrene	87 ± 2	87
Pyrene	95 ± 3	94

Cylinders of carbon dioxide mixed with methanol and other common modifiers are available commercially for chromatographic use. The effect of a methanol modifier in carbon dioxide on capacity factors is illustrated in Figs 7.17 and 7.18. Chromatography was carried out at 60 °C using a column

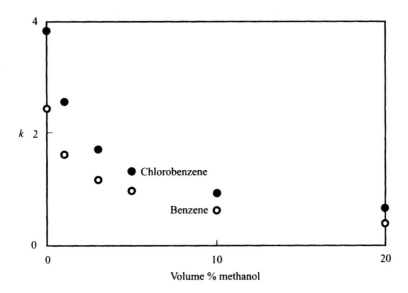

Fig. 7.17 Variation of capacity factor, k, with modifier concentration.

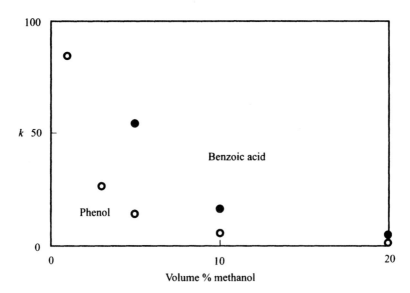

Fig. 7.18 Variation of capacity factor, k, with modifier concentration for two molecules capable of hydrogen bonding.

packed with silica bonded with aminopropyl groups (Heaton *et al.* 1994). The modifier concentration is described, as is typical in analytical chromatography, as the volume per cent as it leaves the pumps. Figure 7.17 shows that methanol reduces the retention of even a non-polar, but polarizable, molecule like benzene and has a greater effect on the more polar molecule chlorobenzene. For even more polar molecules that are capable of hydrogen bonding, a greater effect is seen. In fact phenol and benzoic acid could not be eluted before 60 minutes on this column, and benzoic acid could not be eluted within this time even after the addition of 3 per cent methanol by volume.

The initial rationale behind the use of modifiers was to increase solubility in the fluids, but it is now recognized that this is not the entire explanation of their effect. It is also thought that the modifier may adsorb on to unbonded sites on the silica, and it is observed that improved peak shapes with less tailing are obtained in some cases if modifiers are used. This may be the explanation of the rapid fall in capacity factor, observed in Figs 7.17 and 7.18, with the addition of small amounts of modifier. Thus in chromatography, as in extraction, some of the advantage in the use of modifiers arises from surface effects.

7.4 Simulated moving bed chromatography

The theoretical situation is first considered in which one compound is being injected into a column and in which the stationary phase is no longer stationary. The mobile phase is still considered to move relative to the stationary phase with an average velocity of v_0, but the stationary phase is considered to move in the opposite direction with a velocity of $v_0/(1 + k)$. Unlike the chromatography previously considered, injection of the compound is considered to be continuous and the injection point stationary. The column consists of a number of theoretical plates of height h, and the methods of Section 7.1.2 are used to consider how the compound will move within the column. For simplicity it is considered that r, the fraction of material that moves out of each plate in time δt, is equal to 0.5 and so $1 - r$ will also equal 0.5. Thus the content of each plate arising from what is already in the column after each interval δt will be half the sum of what was in the same plate and in the previous plate at the beginning of the time interval. The relationship between $r = 0.5$ and δt, eqn (7.28), now gives

$$\delta t = 0.5h(1 + k)/v_0 \qquad (7.58)$$

In addition, material is being continuously injected into the column, and as the stationary phase moves, the plate number in which injection will occur will rise. From eqn (7.58) it can readily be calculated that the stationary phase will move a distance h in time $2\delta t$, and so injection will move up one plate in this time. It is considered that a mass $0.5m_0$ of material is injected into a given plate in time δt. Calculations were carried out on this basis and the results are shown in Fig. 7.19, with the crosses giving the quantities in each plate at $4\delta t$, the open circles at $8\delta t$, and the filled circles at $12\delta t$. The peak is rising in area as more material is being injected, but it stays stationary and maintains approximately the same width at half height.

However, to carry out a separation, at least two different compounds will be injected with different capacity factors, and it is also necessary to recover the separated products. To achieve this, the velocity of the stationary phase, v_{st}, is made slightly less than the velocity of component 2, the most retained component, at the injection point, which is itself less than the velocity of

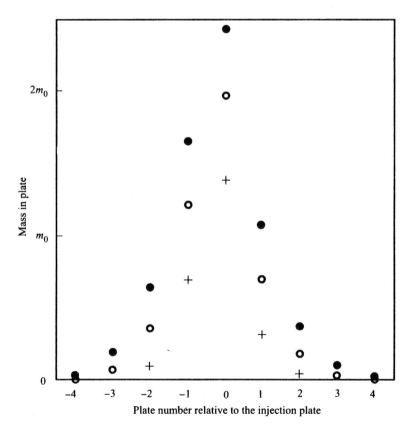

Fig. 7.19 Calculations of peak shape for moving bed chromatography obtained from plate theory, with the crosses giving the quantities in each plate at $4\delta t$, the open circles at $8\delta t$, and the filled circles at $12\delta t$.

component 3, i.e.

$$v_{st} < \frac{v_{0,1}}{1 + k_2} < \frac{v_{0,1}}{1 + k_3} \tag{7.59}$$

where $v_{0,1}$ is the velocity of the mobile phase at the injection point. Both chromatographic peaks then progress slowly with respect to laboratory-fixed coordinates and separate. The situation is illustrated in Fig. 7.20.

At a point further along the column, fluid is bled out so that the mobile phase fluid velocity is reduced to a value $v_{0,2}$ such that

$$v_{st} = \frac{v_{0,2}}{1 + k_2} < \frac{v_{0,2}}{1 + k_3} \tag{7.60}$$

The peak containing component 2 then becomes stationary and component 2 is recovered in the bleed, with component 3 progressing along the column At a later position a further bleed reduces the mobile phase velocity to $v_{0,3}$ such that

$$v_{st} = \frac{v_{0,3}}{1 + k_3} \tag{7.61}$$

and component 3 becomes stationary and is recovered. Component 3 does not progress further and the column can therefore be joined up into a loop, as shown schematically in Fig. 7.21. Fluid must be carried in at the injection point, perhaps as a solvent for the mixture, to make up for that lost in the bleeds. These arguments can be extended to multicomponent mixtures.

In practice, of course, it is difficult to make the stationary phase move and the system is made to work by rotating the injection and recovery points in the same direction as the mobile phase, but more slowly. The method is then known as *simulated moving bed* chromatography. The circular column is

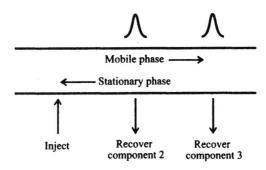

Fig. 7.20 Schematic illustration of moving bed chromatography.

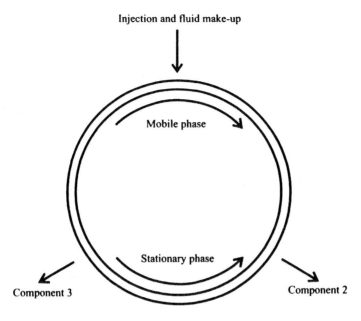

Injection and fluid make-up

Mobile phase

Stationary phase

Component 3

Component 2

Fig. 7.21   Schematic illustration of moving bed chromatography using a closed loop.

divided into a series of zones, which actually contain a number of plates. The injection and recovery points are moved by switching between valves connected at the junctions between zones. Valves between the zones also control the pressure around the column and ensure the direction of the mobile phase flow. As with normal chromatography, flow rate and loading are adjusted to give maximum production rate for a given required purity. A variant of the system, available in SFC but not in liquid chromatography, is to vary the pressure around the loop to broaden the chromatogram at the point of separation and compress it elsewhere. This gives a greater output for a given size of system (Nicoud *et al.* 1993).

8 *Chemical reactions*

The advantages to be gained from carrying out reactions in supercritical fluids, apart from the environmental advantage, can be classified as follows. Firstly, there is control of phase behaviour to obtain either homogenization or separation as required. Secondly, the increased diffusion coefficients can speed up reactions, both homogeneous and heterogeneous, when diffusion is a controlling factor. Thirdly, there is an enhanced control of conditions, through both pressure and temperature, which can increase control of reaction pathways and products. In addition to these there is anecdotal evidence that reactions in supercritical fluids are cleaner and produce less by-products. There is no theoretical reason for this and time only will tell if this is really true, but perhaps the reaction process is smoother in the less dense conditions of a supercritical fluid.

The background in phase behaviour necessary to explain the first advantage has been covered in Chapters 1 to 3 and that in diffusion in Chapter 4. The third advantage of product control may result from diffusion factors or from differential stabilization of the transition state. To give the background for the latter effect, the application of transition-state theory to supercritical fluids will now be discussed.

8.1 Transition-state theory

8.1.1 Basic theory

Transition-state theory, that is treatment of chemical reactions as steady state processes, first devised by Evans and Polanyi (1935,) has been widely used in modelling the rate coefficients of reactions in both gases and liquids. Eyring's extension defining the volume of activation (Glasstone *et al.* 1941), has also been employed to interpret the pressure variation of rate coefficient data, often with mechanistic application. Activation volume is the particular facet of transition-state theory which has received attention in reactions in supercritical fluids, especially in the immediate vicinity of the solvent gas–liquid critical point.

During a reaction process, the reacting species, A and B, are considered to pass through a potential energy maximum in their transition from reagents to products. The minimum energy pathway on the relevant potential energy surface is termed the *reaction coordinate*, and the molecular configuration corresponding to its zenith is the *activated complex* or *transition state species*, $C^{\ddagger}$. Although the activated complex does not have any stable existence, it is regarded as a species and treated thermodynamically. Transition-state theory considers that this species breaks up,

$$C^{\ddagger} \rightarrow \text{products} \tag{8.1}$$

by a first-order process and describes the rate of formation of one of the immediate products, P, in terms of the amount of the activated complex and an appropriately defined rate coefficient, $k^{\ddagger}$:

$$dn_P/dt = k^{\ddagger} n_{C^{\ddagger}} \tag{8.2}$$

where n_i are numbers of moles. It is assumed throughout, unless stated otherwise, that pressure and temperature is kept constant, to simplify the notation in derivatives such as $dn_P/dt \equiv (\partial n_P/\partial t)_{p,T}$.

Although transition-state theory is usually presented in terms of concentrations, in the context of supercritical fluids it is more convenient to use mole fractions, x_i. This is because changes in concentration are often dominated by volume changes, which are not of direct interest, and which cause many of the expressions to have compressibility terms. As can be easily shown by differentiation, if s is the total number of product molecules formed from the activated complex,

$$dx_P/dt = (1/n)\{1 - (s-1)x_P\}dn_P/dt \tag{8.3}$$

eqn (8.3) becomes

$$dx_P/dt = k^{\ddagger}x_{C^{\ddagger}}\{1 - (s-1)x_P\} \tag{8.4}$$

Thus, if only one product molecule is formed from the activated complex, it becomes true that

$$dx_P/dt = k^{\ddagger}x_{C^{\ddagger}} \tag{8.5}$$

and eqn (8.5) is also *approximately* true in other cases, provided that the mole fraction of any product is small, and dilute conditions of reagents and products will now be assumed throughout.

The theory assumes that the activated complex is formed in a rapid pre-equilibrium with the reagents. For example, for the common case of a bimolecular reaction,

$$A + B \rightleftharpoons C^{\ddagger} \rightarrow \text{products} \tag{8.6}$$

the equilibrium between the reagents, A and B, and the activated complex, $C^{\ddagger}$ is supposed to be governed by an equilibrium constant. In a near-critical fluid, the most convenient approach is to define the equilibrium in terms of fugacities, $f_i/p^{\ominus}$, referred to the standard state of $p^{\ominus}$, typically 1 atmosphere:

$$K_p^{\ddagger\ominus} = \frac{f_{C^{\ddagger}}/p^{\ominus}}{(f_A/p^{\ominus})(f_B/p^{\ominus})} \tag{8.7}$$

$K_p^{\ddagger\ominus}$ is an equilibrium constant for the ideal-gas state and is thus equal to that used for dilute-gas reactions. The fugacities are given, in terms of the mole fractions, x_i, the fugacity coefficients, ϕ_i, and the pressure, p, by

$$f_i = \phi_i p x_i \tag{8.8}$$

The substitution of eqns (8.7) and (8.8) into eqn (8.5) gives the following expression for the rate of increase of the product P:

$$\frac{dx_P}{dt} = k^{\ddagger} K_p^{\ddagger\ominus} x_A x_B \cdot \frac{p}{p^{\ominus}} \cdot \frac{\phi_A \phi_B}{\phi_{C^{\ddagger}}} \tag{8.9}$$

In transition-state theory, the first two factors on the right-hand side of eqn (8.9) are modified (leaving the product unchanged in value), using the arguments of statistical mechanics given in many textbooks in physical chemistry, to produce the following more useful equation;

$$\frac{dx_P}{dt} = \frac{k_B T}{h} \cdot \bar{K}_p^{\ddagger\ominus} x_A x_B \cdot \frac{p}{p^{\ominus}} \cdot \frac{\phi_A \phi_B}{\phi_{C^{\ddagger}}} \tag{8.10}$$

where k_B and h are Boltzmann's and Planck's constants, respectively. The bar above $\bar{K}_p^{\ddagger\ominus}$ indicates the modification of the equilibrium constant: it is called *a pseudo equilibrium constant*, but treated like a normal equilibrium constant.

The rate equation in terms of mole fractions and the measured rate constant, k_x, is

$$dx_P/dt = k_x x_A x_B \tag{8.11}$$

Comparison of eqns (8.10) and (8.11) gives the following expression for the rate coefficient:

$$k_x = \frac{k_B T}{h} \cdot \bar{K}_p^{\ddagger\ominus} \cdot \frac{p}{p^{\ominus}} \cdot \frac{\phi_A \phi_B}{\phi_{C^{\ddagger}}} \tag{8.12}$$

Equation (8.12) thus gives the prediction from transition-state theory for the rate of a reaction in terms appropriate for a supercritical fluid. The rate is seen to depend on:

(1) the pressure, the temperature, and some universal constants;

(2) the equilibrium constant for the activated-complex formation in an ideal gas; and

(3) a ratio of fugacity coefficients, which express the effect of the supercritical medium.

Equation (8.12) can therefore be used to calculate the rate constant, if $\bar{K}_p^{\ddagger\ominus}$ is known from the gas-phase reaction or calculated from statistical mechanics, and the ratio $(\phi_A\phi_B/\phi_{C\ddagger})$ estimated from an equation of state. Such calculations are rare; an early example is the modelling of the dimerization of pure chlorotrifluoroethylene $(T_c = 105.8\,°C)$ and comparison with experimental results at $120\,°C$, $135\,°C$, and $150\,°C$ and at pressures up to 100 bar (Simmons and Mason 1972). Fugacities were calculated from the Redlich–Kwong equation of state and there were differences of 30 per cent between experiment and prediction, which reflects the uncertainties in calculating fugacities.

8.1.2   Gibbs function changes on activation

After taking the natural logarithm of eqn (8.12) and using

$$\Delta\bar{G}_p^{\ddagger\ominus} + RT \ln \bar{K}_p^{\ddagger\ominus} = 0 \tag{8.13}$$

where $\Delta\bar{G}_p^{\ddagger\ominus}$ is the modified standard Gibbs function change for the formation of the activated complex in the ideal gas, we obtain

$$\ln k_x = \ln\left(\frac{k_B T}{h}\right) - \frac{\Delta\bar{G}_p^{\ddagger\ominus}}{RT} + \ln\left(\frac{p}{p^\ominus}\right) + \ln\left(\frac{\phi_A\phi_B}{\phi_{C\ddagger}}\right) \tag{8.14}$$

In order to make calculations from this equation, two of the terms are combined to give a Gibbs function change for the reaction, which is still 'standard' in terms of mole fractions, but which applies to the particular supercritical-fluid conditions used, $\Delta G^{\ddagger}$, where

$$\Delta G^{\ddagger} = \frac{\Delta\bar{G}_p^{\ddagger\ominus}}{RT} - \ln\left(\frac{\phi_A\phi_B}{\phi_{C\ddagger}}\right) \tag{8.15}$$

In terms of $\Delta G^{\ddagger}$, eqn (8.14) becomes

$$\ln k_x = \ln\left(\frac{k_B T}{h}\right) - \frac{\Delta G^{\ddagger}}{RT} + \ln\left(\frac{p}{p^\ominus}\right) \tag{8.16}$$

8.1.3   Partial molar volume changes on activation

Using the general thermodynamic relationships $(\partial G/\partial p)_{T,x} = V$ and

$$RT\left(\frac{\partial \ln \phi_i}{\partial p}\right)_{T,x} = V_i - V_i^\ominus - \frac{RT}{p} \tag{8.17}$$

where V_i and $V_i^{\ominus}$ are partial molar volumes under general and ideal-gas conditions, respectively, and the subscript x indicates that composition is kept constant, eqn (8.14) is transformed, after differentiation with respect to pressure at constant temperature, to

$$RT\left(\frac{\partial \ln k_x}{\partial p}\right)_{T,x} = -\Delta \bar{V}^{\ddagger\ominus} - (V_{C\ddagger} - V_A - V_B) + (V_{C\ddagger}^{\ominus} - V_A^{\ominus} - V_B^{\ominus})$$

(8.18)

The terms on the right-hand side of eqn (8.18) are now considered. $\Delta \bar{V}^{\ddagger\ominus}$ is the volume of activation in an ideal gas, as normally used. $(V_{C\ddagger} - V_A \quad V_B)$ is the rate of change of the volume of the reaction mixture with respect to the amount of activated complex formed in moles at constant pressure and temperature. In view of the large partial molar volumes possible in supercritical fluids, this is often important, especially in the critical region. It can appropriately be given the symbol $\Delta V^{\ddagger}$. $(V_{C\ddagger}^{\ominus} - V_A^{\ominus} - V_B^{\ominus})$ is the volume change for the formation of the activated complex in the ideal gas, arising from the volumes of the three species themselves. It is almost identical to the first term $\Delta \bar{V}^{\ddagger\ominus}$, the difference arising from the modification of the equilibrium constant in transition-state theory, as described in the last section, and indicated by the bar.

If this difference between the first and third terms is ignored and they are cancelled, we obtain the equation

$$RT\left(\frac{\partial \ln k_x}{\partial p}\right)_{T,x} = -\Delta V^{\ddagger}$$

(8.19)

Equation (8.19) can be used to calculate a volume of activation from rate-coefficient data in a supercritical fluid. It gives an activation volume which is the difference between the partial molar volume of the activated complex and those of the reagents and includes both intrinsic and solvent effects. Partial molar volumes have been discussed in Section 2.5.3 and shown to be very large and negative in the critical region. Similarly, in many cases activation volumes in supercritical fluids in the critical region are found to be large and negative, both experimentally and in modelling studies. As an example of the latter, modelling of the dimerization of chlorotrifluoroethylene, discussed earlier (Simmons and Mason 1972) gave a value for $\Delta V^{\ddagger}$ of $-3290 \text{ cm}^3 \text{ mol}^{-1}$ close to the critical density and $14\,^{\circ}\text{C}$ above the critical temperature. These extreme negative values are a reflection of the fact that the density of the medium is changing rapidly with pressure and as a consequence the rate coefficient is increasing rapidly from a gas-like to a liquid-like value, and are not an indication of important effects on reaction behaviour due to the proximity of the critical point.

8.1.4   Density dependence of two competing reactions

The large effects due to rapid changes in density with pressure can be elimin-
ated if two reactions are considered which have the same reagents, but produce
two different products. In this case there will be two rate coefficients, k_x' and k_x'',
and also other thermodynamic quantities which will be indicated by the single
and double primes. The ratio of the rate coefficients will be of importance as
this will indicate the ratio of products that will be obtained. By subtracting two
equations of the form of eqn (8.19), the following expression for this ratio is
obtained:

$$RT\left(\frac{\partial \ln(k_x'/k_x'')}{\partial p}\right)_{T,x} = -\Delta V^{\ddagger\prime} + \Delta V^{\ddagger\prime\prime} \tag{8.20}$$

Because the reagents are the same in both cases eqn (8.20) will simplify to

$$RT\left(\frac{\partial \ln(k_x'/k_x'')}{\partial p}\right)_{T,x} = -V_{C\ddagger}' + V_{C\ddagger}'' \tag{8.21}$$

Using eqn (2.7) for the partial molar volumes and assuming that all reagents,
transition states, and products are at infinite dilution, the following expression
is obtained, after the molar volume terms cancel

$$RT\left(\frac{\partial \ln(k_x'/k_x'')}{\partial p}\right)_{T,x} = \left(\frac{\partial V}{\partial p}\right)_{T,x}\left[\left(\frac{\partial p}{\partial x_{C\ddagger}'}\right)_{T,V}^{\cdot} - \left(\frac{\partial p}{\partial x_{C\ddagger}''}\right)_{T,V}\right] \tag{8.22}$$

The term involving the compressibility, $(\partial V/\partial p)_{T,x}$, can now be divided out
of eqn (8.22) to give

$$RT\left(\frac{\partial \ln(k_x'/k_x'')}{\partial V}\right)_{T,x} = \left(\frac{\partial p}{\partial x_{C\ddagger}'}\right)_{T,V} - \left(\frac{\partial p}{\partial x_{C\ddagger}''}\right) \tag{8.23}$$

Thus the density dependence of the ratio of rate coefficients is related to the
difference in tuning functions for the two transition states. As was discussed
in Sections 2.5.1 and 2.5.2, these tuning functions arise from interactions
between the transition states and the solvent molecules. This effect can give rise
to product control, as is discussed in Section 8.4.2.

8.2   Control of phase behaviour

The most straightforward reason for carrying out chemical reactions in
supercritical fluids is that control of phase behaviour by pressure and temperature may allow reagents or products to be present either in one phase
(homogenization) or two phases (separation), as required. During a reaction,

homogenization may be desirable initially followed by separation later in the process. Phase homogenization can mean that a reaction which would otherwise be heterogeneous can be made homogeneous and therefore more rapid and effective. This is often because the pressures used bring the system closer to, or above, a critical line in the phase diagram. For example, larger concentrations of light gases such as hydrogen and oxygen can be dissolved in the reaction medium. A special case is water, which under supercritical conditions is much less polar and can homogenize substantial amounts of non-polar organic compounds, making them available for reaction. Another example is that products, which under low-pressure conditions would deposit on and 'poison' a catalyst, can sometimes be solvated by a supercritical fluid.

In other reactions phase control is used to separate products. An example would be the catalytic oxidation of a sulphide dissolved in supercritical carbon dioxide, where conditions were chosen so that the more polar sulphoxide product would precipitate from solution. These schemes allow ready separation of the product and may be used to push a chemical equilibrium towards the product. They may also be used to separate an intermediate product, such as a partially oxidized hydrocarbon. Phase control is particularly important in polymerization reactions. This was understood to be the case from the 1940s well before the term supercritical fluid was used, and, as the area is extensively covered elsewhere, it will only be discussed briefly here.

8.2.1 Phase homogenization

The advantage of phase homogenization arises from the fact that the reaction rate is dependent upon the concentrations of the reagents, often certain key reagents, and also sometimes on the catalyst concentration. In certain reactions, supercritical fluids can therefore enhance reaction rates by bringing key reagents and catalysts into the same phase. There are a variety of situations in which this is possible and so two examples are given at the end of this section to indicate the scope. In each case, the methods described in Chapter 2 can be used to investigate the phase behaviour of a system of interest. In particular, a phase envelope, as illustrated in Fig. 2.6 can be readily calculated for a multicomponent mixture using commercially available software. It is sometimes found that when published work on reactions in supercritical fluids is analysed, conditions place the reaction mixture to the left of the critical point in Fig. 2.6, and so the reaction is actually taking place in the liquid phase and can at best be described as near critical. This may not matter if the pressure is high enough so that only one phase exists, but desired effects due to the compressibility of supercritical fluids may not be present.

An example is given here for the Diels–Alder reaction of cyclopentadiene and methyl acrylate in carbon dioxide at 35 °C. Phase envelopes for the same amounts of the reagents but different amounts of carbon dioxide, i.e. different concentrations of reagents, are shown in Fig. 8.1, obtained using the Peng–Robinson equation of state. Critical temperatures of 304 K, 536 K, and 500 K; critical pressures of 74 bar, 37 bar, and 47 bar; and acentric factors of 0.223, 0.292, and 0.195 were used for carbon dioxide, methyl acrylate, and cyclopentadiene, respectively. The unknown binary interaction parameters were set to zero. Some calculations were carried out with the methyl acrylate–CO_2 and the cyclopentadiene–CO_2 interaction parameters set to 0.1. These showed that the general conclusions drawn below are not sensitive to the values of the interaction parameters. It can be seen in Fig. 8.1 that phase envelopes collapse as the concentration is lowered and at infinite dilution would become the vapour pressure curve for pure carbon dioxide. It was found that at the two highest concentrations, the experimental temperature is 60 K and 18 K below the mixture critical temperature, putting the conditions in the liquid region. For the lowest concentration the experimental temperature is 1 K above the mixture critical temperature and the system is thus supercritical, provided the pressure is high enough.

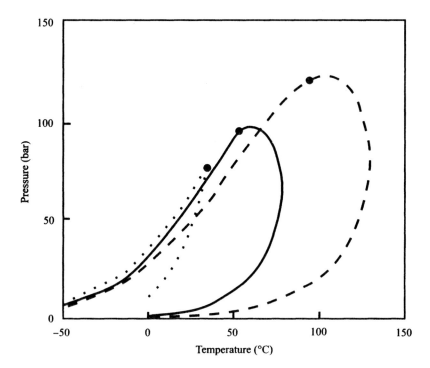

Fig. 8.1 Phase envelopes, calculated from the Peng–Robinson equation of state, for mixtures containing 1 mol of cyclopentadiene, 1.82 mol of methyl acrylate, and various amounts of CO_2: 64.8 mol (continuous line); 17.7 mol (dashed line); and 617.2 mol (dotted line).

As explained earlier, the difference between experiments which are carried out at temperatures above those of the phase envelope, as opposed to experiments at lower temperatures, is one of the degree of compressibility. Figure 8.2 shows the dependency of the density on pressure at 35 °C for the three compositions studied, and shows that the mixture which is more dilute in reagents has a greater compressibility. Changing the pressure from 80 bar (a lower limit which will not penetrate the two-phase envelope for any of the compositions) to 300 bar changes the density, from $500 \, kg \, m^{-3}$ to nearly $1000 \, kg \, m^{-3}$ at the lowest concentration, but only changes the density from $750 \, kg \, m^{-3}$ to $850 \, kg \, m^{-3}$ at the highest concentration. These curves were calculated using the Peng–Robinson equation of state and again give the trends rather than accurate absolute values. They show that a greater range of densities can be studied at lower concentrations.

Two examples are now given to illustrate principles involved in phase homogenization by supercritical fluids, beginning with the solution of a solid catalyst. A patent by Kramer and Leder (1975) describes a C_4–C_{12} normal or isoparaffinic hydrocarbon isomerization process employing supercritical carbon dioxide. At the temperatures and pressures used (up to 80 °C, and up to 170 bar) the preferred Friedel–Crafts catalyst, aluminium bromide, was soluble, and used homogeneously in the concentration range 0.1 to 0.5 M. Addition of soluble gaseous hydrogen to the medium promoted isomerization over cracking. Conversions were similar to heterogeneous liquid–solid reactions, but with greater selectivity for the isomerization products. The effect

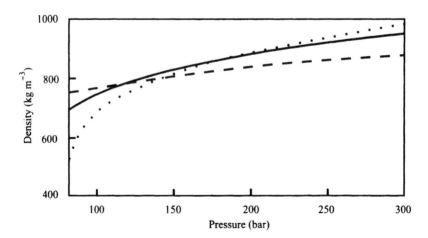

Fig. 8.2 Densities at 35 °C for a range of pressure, calculated from the Peng–Robinson equation of state, for mixtures containing 1 mol of cyclopentadiene, 1.82 mol of methyl acrylate, and various amounts of CO_2: 64.8 mol (continuous line); 17.7 mol (dashed line); and 617.2 mol (dotted line).

of phase on reaction rate was not reported, but a further advantage of super-critical conditions is the removal of reaction products or spent catalyst by raising the temperature or lowering the pressure.

A second example is where light gases can be dissolved readily in supercritical fluids. The reagents hydrogen and nitrogen are fully miscible with carbon dioxide and xenon under supercritical conditions. Thus, for a given pressure, the reagents can be at a much higher concentration than is possible in a liquid. The effect allows the formation of metal complexes which would be difficult to obtain otherwise (Howdle *et al.* 1990). For example, carbon monoxide molecules in the rhenium complex [cpRe(CO)$_3$], where cp is cyclopentadiene, could be successively substituted by nitrogen to form [cpRe(CO)$_{3-x}$(N$_2$)$_x$], where $x = 1$, 2, or 3. The complexes formed were stable enough to isolate and had resistance to reverse substitution by carbon monoxide.

8.2.2   Solubilization of catalyst poisons

Supercritical fluids can also be used to solvate and remove unwanted by-products, which would otherwise poison or foul a reaction catalyst. They can also be used to solvate various interfering agents and achieve the maintenance or reactivation of heterogeneous catalysts. These are susceptible to deactivation under liquid and gaseous conditions by a number of processes, such as coking, fouling, and poisoning. Tiltscher *et al.* (1984) considered the catalysis of double-bond isomerization of 1-hexene on γ-Al$_2$O$_3$, with 2-chlorohexane as co-catalyst. The C$_{12}$ to C$_{20}$ oligomerization by-products are of low volatility and, under gaseous conditions, coke the catalyst as they do not desorb or separate from its surface as readily as the hexenes. In liquid hexene, removal of low-volatility species is mass-transfer limited: periodic change of the system to supercritical conditions was used to dissolve the oligomers and facilitate their transport from the catalyst surface. Catalyst lifetime was extended threefold in this way.

In another study by the same authors, 1,4-diisopropylbenzene dispropor-tionates to cumene and 1,3,5-triisopropylbenzene, in benzene, on a zeolyte catalyst, the micropores of which are large enough to house reagents and products; however, 1,3,5-triisopropylbenzene cokes the zeolite. Pentane was added to lower the mixture critical temperature and at the higher pressures used under supercritical conditions, dissolved the 1,3,5-triisopropylbenzene, so extending the catalyst lifetime.

8.2.3   Separation of products

In a near-critical or supercritical fluid, products or by-products of a reaction are separated by exploiting the inherent phase characteristics of the fluid system. This may be done, for example, by forming a product which is insoluble under reaction conditions or employing a pressure or temperature change at the conclusion of a reaction to the same end. The product(s) may be precipitated for the purpose of recovery, to obtain an intermediate, or, in a reaction at equilibrium, to drive the reaction forward. Precipitation of the products and catalyst was used in the study of the isomerization of paraffins, described in Section 8.2.1 (Kramer and Leder 1975). A knowledge of the phase behaviour of the reacting system is therefore necessary for the design of separation conditions.

The technique applies equally when the fluid is a reacting system; for example, free radical polymerization of ethene is, under certain conditions, subject to spontaneous precipitation of higher molecular weight polymers. This has been understood for some time and precipitation may be deliberately induced by partial depressurization of the reaction medium (Krase and Lawrence 1946). The precipitation threshold may be governed by molar mass, or by other properties such as oligomer architecture and the ability to crystallize.

8.3 Diffusion effects

Some reactions are controlled by the rate of diffusion of reagents towards each other. Examples are heterogeneous reactions, such as enzymatic reactions, and very fast homogeneous reactions, such as some free radical reactions. In supercritical fluids diffusion coefficients are generally higher than in typical liquids and this can enhance reaction rates. A special case, again, is water, where the breakdown of the hydrogen-bonded structure under supercritical conditions can allow much more rapid diffusion and reaction. Although generally higher than in liquids, binary diffusion coefficients can, conversely, exhibit an 'anomalous' lowering in the critical region and may thus cause a reduced reaction rate or give rise to product control

8.3.1 Homogeneous reactions

For some rapid reactions, the rate of reaction is controlled by the rate of diffusion of the two reagent molecules. In some cases, as pressure increases in a supercritical fluid, a reaction may pass from activation control to diffusion control. This has been demonstrated, for example, in investigations by Hippler *et al.* (1984) into the recombination of bromine atoms in various bath gases. by measuring quantum yields following laser photolysis.

In the simplest analysis of diffusion control, the second order rate coefficient, k_x, is given by the Smoluschowski equation:

$$k_x = 4\pi N_A r \phi (D_A + D_B) \tag{8.24}$$

where D_A and D_B are the binary diffusion coefficients of the reagents in the fluid, r is the distance of approach necessary for reaction, and ϕ is a statistical or steric parameter of less than unity which takes into account, for example, the necessary relative orientation of the two molecules for the reaction to take place. The diffusivity of a dilute solute in a supercritical fluid, somewhat removed from the critical point, is typically an order of magnitude greater than in liquid solvents at comparable temperatures. The behaviour of reactions is sometimes interpreted in terms of changes in the viscosity of the fluid. Inherent in these interpretations is an assumption of the validity of the Stokes–Einstein equation relating the diffusion coefficient to the fluid viscosity, and they are therefore equivalent to discussions in terms of diffusion. The appropriate equation obtained is

$$k_x = 8RT/3\eta \qquad (8.25)$$

where η is the coefficient of viscosity of the medium. Good agreement between experiment and the predictions of eqn (8.25) has been obtained in supercritical fluids (Brennecke 1993). According to eqn (8.25), all diffusion-controlled second-order rate coefficients are the same in a particular solvent.

Diffusion is also important for unimolecular fission. Thus, radical initiators under supercritical fluid conditions are able to escape more readily from solvent cages, and the rate coefficient for the initiation process is markedly increased. Processes propagated by free radicals, such as polymerization, are also rate enhanced. For unimolecular decompositions, initiated thermally or by light, decreasing the diffusion coefficient by increasing density can decrease the decomposition rate. This is ascribed to slow diffusion out of a 'cage' of solvent molecules giving rise to an increased rate of the 'geminate' recombination of two molecular fragments formed by decomposition.

Perhaps the most visible demonstration of diffusion control in supercritical fluids is the existence of typical diffusion flames in supercritical water up to 2000 bar (Schilling and Franck 1988). The high diffusion coefficients in supercritical water, discussed in Section 4.2.7, combined with the non-polar nature of water under these conditions, discussed in Section 1.4.1, explain this experiment. These features also explain the fact that organic compounds can be rapidly and almost completely (99.99 per cent) converted by molecular oxygen to benign small species: CO_2, H_2O, N_2, Cl^-, SO_4^-, etc. (Modell 1982). This has given rise to widespread research into processes for the safe destruction of toxic materials. Residence times of around one minute are usually required, compared with 30 minutes for the wet oxidation processes that are carried out at below 300 °C. Although diffusion in supercritical fluids is often desirably faster than in liquids, if a reaction is being compared with a gas-phase process, diffusivities may appear slow. Oxidation rate coefficients of simple molecules in supercritical water, studied by Hellig and Tester (1987), are reduced in comparison with gas-phase combustion. For example, the rates of oxidation of carbon monoxide are substantially lower than those predicted by gas-phase models, and the proportion of hydrogen produced by the concurrent water–gas shift reaction is unexpectedly high. The differences are explained in terms of lower diffusivities compared with the gas phase.

In the region of the critical point, diffusion coefficients can fall for finite concentrations, as discussed in Section 4.2.5. The behaviour of reactions in the critical region can therefore be discussed using eqns (8.23) and (4.34). This type of approach will be used in Section 8.4.1. However, in a more integrated approach, the methods of non-equilibrium thermodynamics (de Groot and Mazur 1962) can be used as a basis for discussion of what effects can be expected on both the rates, including diffusion-controlled rates, and equilibrium positions of chemical reactions due to the proximity of a critical point. These arguments have been reviewed and applied to the discussion of a number of experimental studies by Greer (1985).

8.3.2 Heterogeneous reactions

Very often catalytic heterogeneous reactions are controlled by the rate of diffusion to the catalyst surface, and enhanced diffusion rates in supercritical fluids can be an advantage. The speed can be increased and, in the case of a continuous process, residence times in the reactor can be reduced and smaller vessels used for a given output. The reduction of reactor size can offset the extra expense in using vessels designed for higher pressures.

An important category of such reactions are enzymatic conversions, as enzymes are insoluble in normal solvents and the use of non-toxic carbon dioxide can be an advantage. Diffusion is not only enhanced in the bulk fluid, but also diffusion within the pores of particles containing enzyme molecules is more rapid, and the swelling of such particles is enhanced, compared with that in liquid solvents. Although some enzymes have been shown to be thermodynamically unstable when exposed to supercritical fluid conditions, many of those which are stable retain their catalytic activity. Biochemical substrates are denatured or thermally degraded at moderate temperatures, and enzymes themselves are only active over a relatively narrow temperature range, typically 10 to 80 °C. This enforces limits on the range of suitable supercritical fluids and of these, carbon dioxide has been the most favoured. A large number of enzymatic reactions in carbon dioxide have been studied, many of these being transesterification reactions using lipase. These are important reactions in the processing of lipids, as triglycerides typically contain a mixture of fatty acids, and conversion to the methyl or ethyl esters allows fatty acids of nutritional or pharmaceutical interest to be separated. The presence and concentration of water and alcohols in the fluid often has a large effect on the reaction rate. It is thought that these compounds affect the conformation of the enzyme, and sometimes the substrate, by controlling the relative amounts of hydrogen bonding that the enzyme has within itself and with molecules in the fluid.

8.4   Product control

In complex reactions, with a range of possible products, some of the reaction steps may be diffusion-controlled, while others are not. Control of the diffusion coefficient by density or in the critical region can change the relative rates of the reaction steps. Thus the course of the reaction, and in some cases, the ratio of the products obtained can be controlled. An alternative mechanism of product control is that the transition state leading to a particular product may be stabilized by solvation more than the transition state leading to another product.

8.4.1   Product control due to diffusion

In a complex reaction scheme, where only some of the steps are diffusion controlled, the course of the reaction can be controlled by controlling diffusion coefficients. This can be done in two ways: away from the critical region, because of the inverse relationship of density and diffusion coefficient; and in the critical region by using the critical 'anomaly'. The former of these is most straightforward and is discussed first.

The archetypal example here is polymerization. The effect of diffusion control on overall rate and the rates of the different steps is first discussed. A simplified equation, sufficient for a qualitative discussion, for the instantaneous rate of polymer formation is as follows:

$$d[x_{polymer}]/dt = (\varepsilon x_I k_i/k_t)^{1/2} k_p x_M \qquad (8.26)$$

where ε is the initiator efficiency, x_I and x_M, the mole fractions of initiator and monomer, and k_i, k_p and k_t the rate coefficients for the initiation, propagation, and termination processes, respectively. The factor $k_p/k_t^{1/2}$, which gives the pressure dependence of the reaction if its initiation is pressure independent, is found to increase with increasing pressure. This is readily explained by termination steps becoming diffusion controlled at much lower pressures than propagation steps, because the former, in general, involve large polymer radicals, whereas the latter involve smaller monomer molecules as one of the reacting species. However, initiation may also be affected. It is suggested that diffusion control is associated with the formation of solvent *cages*, indicated below by curly brackets, around reactant molecules:

$$AB \rightleftharpoons \{A \cdot \ \cdot B\} \rightarrow A \cdot + \cdot B \qquad (8.27)$$

When the molecule concerned is a radical initiator this has implications for the efficiency of radical production, related to cage escape. Under supercritical conditions, as the density increases, diffusion out of the cage becomes slower and geminate recombination within the cage increases in importance.

Critical region effects on the diffusion coefficient, described in Section 4.2.5, can affect product ratios. This is probably the explanation of the increase of the photo-Fries enolization products near the critical density in the photochemical reaction of 1-naphthyl acetate with 2-propanol in supercritical carbon dioxide (Andrew *et al.* 1995). The photo-Fries enolization products are produced within the solvent cage and diffusion out of the cage is necessary to produce the other product, 1-naphthol. Outside the critical region the ratio of photo-Fries enolization products to naphthol is around 4, but this increases to more than 12 near the critical density.

8.4.2 Product control due to effects on the transition state

In a situation where reagents can react to form two or more products, there is experimental evidence and considerations based on transition state theory which indicate that the ratios of the product yields can be usefully controlled by adjusting the conditions in a supercritical fluid. We consider a reaction in which two products are obtained, the appropriate parameters being indicated by single and double primes, and use eqn (8.23), obtained in Section 8.1.4, in which the density dependence of the ratio of rate coefficients is related to the difference in tuning functions for the two transition states. These tuning functions were discussed in Sections 2.5.1 and 2.5.2, and it was shown that they arise from intermolecular interactions between the solute molecules, in this case the transition states, and the solvent molecules. These effects are now discussed qualitatively using the van der Waals equation of state.

In Section 2.3.1 an expression, eqn (2.10), for the tuning functions in terms of van der Waals parameters was obtained, which is now substituted into eqn (8.23) to obtain at infinite dilution

$$RT\left(\frac{\partial \ln(k'_x/k''_x)}{\partial V}\right)_{T,x} = \frac{RT(b'_2 - b''_2)}{(V - b_1)^2} - \frac{2(a'_{12} - a''_{12})}{V^2} \tag{8.28}$$

It is now convenient to define the quantities, β and γ, by

$$\beta = (b'_2 - b''_2) \tag{8.29}$$

$$\gamma = \frac{2(a'_{12} - a''_{12})}{R(b'_2 - b''_2)} \tag{8.30}$$

whereupon eqn (8.28) becomes

$$\left(\frac{\partial \ln(k'_x/k''_x)}{\partial V}\right)_{T,x} = \beta\left(\frac{1}{(V-b_1)^2} - \frac{\gamma}{TV^2}\right) \tag{8.31}$$

Integrating eqn (8.31) and using the boundary condition that $(k'_x/k''_x) \rightarrow (k'_x/k''_x)_0$ as $V \rightarrow \infty$ and taking exponentials, we obtain

$$(k'_x/k''_x) = (k'_x/k''_x)_0 \exp\{-\beta[1/(V-b_1) - \gamma/TV]\} \tag{8.32}$$

which is an equation with which experimental data can be compared. It has three unknown parameters: the rate coefficient ratio under perfect gas conditions and two parameters related to the differences in the a and b van der Waals parameters for the two transition states. The latter two parameters are expected to be independent of density. A value of b_1 can be obtained from the critical density of the solvent.

By choice of the β and γ parameters various forms of the behaviour of (k'_x/k''_x) with respect to density can be theoretically obtained, including the presence of a maximum. An experimental study on the Diels–Alder reaction between cyclopenta-1,3-diene and methyl acrylate was carried out (Clifford *et al.* 1997*b*). The reaction produces two stereoisomeric products: the *endo* and *exo* forms. Experiments were carried out at the lowest concentrations described in Section 8.2.1, to ensure that conditions were supercritical, which also means that the limit of infinite dilution assumed in deriving the equations given above will be more closely approached. Experiments were carried out at 35 °C and 40 °C and at a range of pressures between 80 bar and 270 bar, and data for the ratio of *endo* to *exo* product are plotted against density in Fig. 8.3. For this purpose, the density of the solution was assumed to be that of pure CO_2 and was calculated from the Span and Wagner equation of state (1996). The product ratio is seen to go through a maximum, although the range of values of the ratio is no different than can be obtained by changing the solvent in the liquid-phase reaction (Berson *et al.* 1962) . However, by controlling the density, this range can be obtained in a single solvent. The maximum in the product ratio does not occur at the critical density.

The curves shown in Fig. 8.3 are theoretical curves obtained from eqn (8.32) using a value for b_1 for CO_2 of $31.3 \times 10^{-6} \, \mathrm{m^3 \, mol^{-1}}$, obtained from its critical density. The other three parameters in eqn (8.32) were obtained by fitting the experimental data at 40 °C, and these same three parameters were used to produce the theoretical curve at 35 °C. The values of the parameters found by fitting were $(k'_x/k''_x)_0 = 1.71$, $\beta = 26 \times 10^{-6} \, \mathrm{m^3 \, mol^{-1}}$, and $\gamma = 1239 \, \mathrm{K}$. From these, the differences in the van der Waals parameters for the interactions of the two transition states with CO_2 are found to be $(a'_{12} - a''_{12}) = 0.134 \, \mathrm{J \, m^3 \, mol^{-2}}$ and $(b'_2 - b''_2) = 26 \times 10^{-6} \, \mathrm{m^3}$. For the naphthalene–$CO_2$

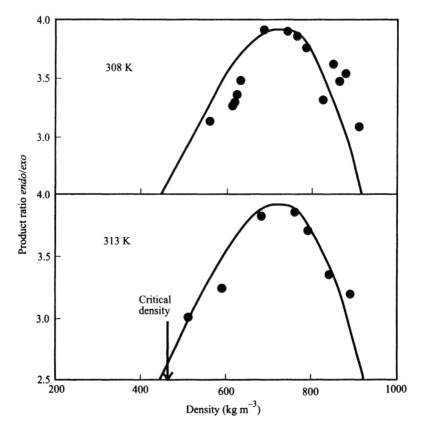

Fig. 8.3 Experimental data, shown as points, for the ratio of *endo* to *exo* product from the reaction of methyl acrylate and cyclopentadiene in carbon dioxide at 308 K and 313 K, with the curves showing predictions from eqn (8.32).

system (naphthalene being a molecule of approximately the same size as the transition states), $a_{12} = 1.03 \, \mathrm{J \, m^3 \, mol^{-2}}$ and $b_2 = 189 \times 10^{-6} \, \mathrm{m^3}$. These values are almost an order of magnitude greater than the differences obtained for the corresponding parameters for the two transition states. Thus the parameters calculated from the experimental results are reasonable in magnitude. The rising part of the curves in Fig. 8.3 are associated with attractive forces and the steeper falling parts with repulsion.

The maximum observed in the density variation of the ratio of the *endo* to *exo* product from this Diels–Alder reaction does not occur at the critical density, and thus is not associated with clustering in the critical region. Instead, it appears that the maximum is associated with tuning of the average distance between the solute molecules and the transition states as the pressure and density vary. It thus appears that in a compressible medium such as a super-critical fluid it may often be possible, by controlling the density, to adjust the mean distance between a solvent molecule and a more massive solute species within a range where their mutual pair potential function is varying significantly. The solvation energy of the solute can therefore be controlled by density and this will cause variations in a physical or chemical equilibrium and, if the species is a transition state, the rate of a chemical reaction.

8.4.3   Control of equilibria

Reaction rates have been the main concern of this chapter, but many of the effects on rate coefficients will also apply to their ratio, the equilibrium constant, expressed here in terms of mole fractions, K_x. Using methods analogous to those used for developing the transition state theory, the following relationship can be obtained

$$\left(\frac{\partial \ln K_x}{\partial p}\right)_{T,x} = -\Delta V \tag{8.33}$$

where ΔV is the difference between the partial molar volumes of the products minus those of the reagents, multiplied by their stoichiometries. Thus the behaviour of partial molar volumes in supercritical fluids may cause interesting behaviour. If there is the same number of product molecules as reagent molecules, the molar volume terms in eqn (2.7) will cancel, and the rate of change of equilibrium constant with density becomes equal to differences in tuning functions. A number of studies have been made and one example is described here: the tautomerization between 2-hydroxypypiridine and 2-pyridone in 1,1-difluoroethane and propane (Peck *et al.* 1989). The reaction was monitored *in situ* spectroscopically, at reagent concentrations causing negligible self-association of 2-pyridone. The equilibrium constant was shown to be a strong function of pressure for both solvents in their respective critical regions, with stronger variation in polar 1,1-difluoroethane.

References

Andrew, D., Des Islet, B. T., Margaritis, A., and Weedon, A. C. (1995). *Journal of the American Chemical Society*, **117**, 6132.

Angus, S., Armstrong, B., and de Reuck, K. M. (1976). *International thermodynamic tables of the fluid state: Volume 3 carbon dioxide*. Pergamon, Oxford.

Aris, R. (1956). *Proceedings of the Royal Society of London*, **A235**, 67.

Ashraf, S., Bartle, K. D., Clifford, A. A., Raynor, M. W., and Shilstone, G. F. (1992). *Analyst*, **117**, 1697.

Bartle, K. D., Clifford, A. A., Kithinji, K. P., and Shilstone, G. F. (1988). *Journal of the Chemical Society, Faraday Transactions I*, **84**, 4487.

Bartle, K. D., Boddington, T., Clifford, A. A., and Shilstone, G. F. (1989). *Journal of Chromatography*, **471**, 347.

Bartle, K. D., Clifford, A. A., and Shilstone, G. F. (1990a). *Journal of Supercritical Fluids*, **3**, 143.

Bartle, K. D., Clifford, A. A., and Jafar, S. A. (1990b). *Journal of the Chemical Society, Faraday Transactions*, **86**, 855.

Bartle, K. D., Clifford, A. A., Jafar, S. A., and Shilstone, G. F. (1991a). *Journal of Physical and Chemical Reference Data*, **20**, 713.

Bartle, K. D., Boddington, T., Clifford, A. A., Cotton, N. J., and Dowle, C. J. (1991b). *Analytical Chemistry*, **63**, 2371.

Bartle, K. D., Clifford, A. A., and Shilstone, G. F. (1992). *Journal of Supercritical Fluids*, **5**, 220.

Bartle, K. D., Bevan, C. D., Clifford, A. A., Jafar, S. A., Malak, N., and Verrall, M. S. (1995). *Journal of Chromatography*, **A697**, 579.

Basile, A. (1997). MSc Thesis, University of Leeds.

Berson, J. A., Hamlet, Z., and Mueller, J. (1962). *Journal of the American Chemical Society*, **84**, 297.

Brennecke, J. F. (1993). In *Supercritical fluid engineering science*, ACS Symposium Series No. 514 (ed. E. Kiran and J. F. Brennecke), Ch. 16. American Chemical Society, Washington, DC.

Brünner, E., Hültenschmidt, W., and Schlichthärle, J. (1987). *Journal of Chemical Thermodynamics*, **19**, 273.

Cagniard de la Tour, C. (1822). *Annales de Chimie et de Physique*, **22**, 127.

Caralp, M. H. M., Clifford, A. A., and Coleby, S. E. (1993). In *Extraction of natural products using near-critical solvents* (ed. M. B. King and T. R. Bott), Ch. 3. Blackie, Glasgow.

Carroll, J. L. (1991), MSc Thesis, University of Leeds.

Carslaw, H. S., and Jaeger, J. C. (1959). *Conduction of heat in solids*. Clarendon, Oxford.

Clifford, A. A. and Coleby, S. E. (1991). *Proceedings of the Royal Society of London,* **A433**, 63.

Clifford A. A., Burford, M. D., Hawthorne, S. B., Langenfield, J. D., and Miller, D. J. (1995). *Journal of the Chemical Society, Faraday Transactions*, **91**, 1333.

Clifford, A. A., Bartle, K. D., Gélébart, I., and Zhu, S. (1997a) *Journal of Chromatography*, A785, 395.

Clifford, A. A., Pople, K., Gaskill, W. J., Bartle, K. D., and Rayner, C. M. (1997b). *Chemical Communications*, 595.

Cochran, H. D., and Lee, L. L. (1989). In *Supercritical fluid science and technology,* ACS *Symposium Series, No. 406* (ed. K. P. Johnstone and J. M. L. Penninger), pp. 27–38. American Chemical Society, Washington.

Cowey, C. M., Bartle, K. D., Burford, M. D., Clifford, A. A., and Zhu, S. (1995). *Journal of Chemical and Engineering Data*, **40**, 1217.

Crank, J. (1975). *The mathematics of diffusion*. Clarendon, Oxford.

de Groot, S. R., and Mazur, P. (1962). *Non-equilibrium thermodynamics*. North Holland, Amsterdam.

Evans, M. G., and Polanyi, M. (1935). *Transaction of the Faraday Society*, **31**, 875.

Feist, R., and Schneider, G. M. (1982). *Separation Science and Technology*, **17**, 261.

Fjeldsted, J. C., Jackson, W. P., Peaden, P. A., and Lee, M. L. (1983). *Journal of Chromatographic Science*, **21**, 222.

Glasstone, S., Laidler, K. J., and Eyring, H. (1941). *The theory of rate processes* (1st edn). McGraw Hill, New York.

Golay, M. J. E. (1958). In *Gas chromatography 1958* (ed. D. H. Desty), p. 36. Butterworths, London.

Greer, S. C. (1985). *Physical Review*, **A31**, 3240.

Haar, L., Gallagher, J. S., and Kell, G. S. (1984). *NBS/NRC Steam tables.* Hemisphere, Washington.

Hawthorne, S. B., Galy, A. B., Schmitt, V. O., and Miller, D. J. (1995). *Analytical Chemistry*, **67**, 2723.

Heaton, D. M., Bartle, K. D., Clifford, A. A., Klee, M. S., and Berger, T. A. (1994). *Analytical Chemistry*, **66**, 4253.

Hellig, R. K. and Tester, J. W. (1987). *Journal of Energy and Fuels*, **1**, 417.

Hippler, H., Schubert, V., and Troe, J. (1984). *Journal of Chemical Physics*, **81**, 3931.

Howdle, S. M., Healey, M. A., and Poliakoff, M. (1990). *Journal of the American Chemical Society*, **112**, 4804.

Hunter, E., and Richards, R. B. (1948). *US* Patent 2 457 238.

Johns, A. I., Raschid, S., Watson, J. T. R., and Clifford, A. A. (1986). *Journal of the Chemical Society, Faraday Transactions I*, **82**, 2235.

Johnston, K. P., Peck, D. G., and Kim, S. (1989). *Industrial and Engineering Chemistry Research*, **28**, 1115.

Kirkwood, J. G., and Buff, F. P. (1951). *The Journal of Chemical Physics*, **19**, 774.

Knapp, H., Döring, R., Oellrich, L., Plöcker, U., and Prausnitz, J. M. (1982). *Chemistry Data Series*, Vol VI. DECHEMA, Frankfurt.

Knox, J. H., and McLaren, L. (1964). *Analytical Chemistry*, **36**, 1477.

Kramer, A., and Thodos, G. (1988). *Journal of Chemical and Engineering Data*, **33**, 230.

Kramer, G. M., and Leder, F. (1975). US Patent, 3 889 945.

Krase, N. W., and Lawrence, A. E. (1946). US Patent, 2 396 791.

Kurnik, R. T., and Reid, R. C. (1982). *Fluid Phase Equilibria*, **8**, 93.

Lauer, H. H., McManigill, D., and Board, R. D. (1983). *Analytical Chemistry*, **55**, 1370.

Lee, M. L., and Markides, K. E. (1990). *Analytical supercritical fluid chromatography and extraction*. Chromatographic Conferences, Inc., Provo, Utah.

Mack, E. (1925). *Journal of the American Chemical Society*, **47**, 2468.

Maitland, G. C., Rigby, M., Smith, E. B., and Wakeham, W. A. (1981). *Intermolecular forces*. Clarendon, Oxford.

Marshall, W. L., and Franck, E. U. (1981). *Journal of Physical and Chemical Reference Data*, **10**, 295.

McHugh, M. A., and Krukonis, V. J. (1994). *Supercritical fluid extraction*, (2nd edn). Butterworth–Heinemann, Boston.

McHugh, M. A., and Paulitis, M. E. (1980). *Journal of Chemical and Engineering Data*, **25**, 326.

Miller, D. J., Hawthorne, S. B., Clifford, A. A., and Zhu, S. (1996). *Journal of Chemical and Engineering Data*, **41**, 779.

Mills, N. J. (1986). *Plastics – Microstructure, properties and applications*. Edward Arnold, London.

Modell, M. (1982). US Patent, 4 338 199.

Najour, G. C., and King, A. D. (1966). *Journal of Chemical Physics*, **45**, 1915.

Nicoud, R.-M., Perrut, M., and Hotier, G. (1993). World patent, WO 93/22022.

Nilsson, W. B. (1996). In *Supercritical fluid technology in oil and lipid chemistry* (ed. J. W. King and G. R. List), Ch. 8. American Oil Chemists' Society, Champaign, Illinois.

Özcan, A. S., Clifford, A. A., Bartle, K. D., and Lewis, D. M. (1997). *Journal of Chemical and Engineering Data*, **42**, 590.

Page, S. H., Sumpter, S. R., and Lee, M. L. (1992). *Journal of Microcolumn Separations*, **4**, 91.

Peck, D. G., Mehta, A. J., and Johnston, K. P. (1989). *Journal of Physical Chemistry*, **93**, 4297.

Peng, D. Y., and Robinson, D. B. (1976). *Industrial and Engineering Chemistry Fundamentals*, **15**, 59.

Pitzer, K. S. (1955). *Journal of the American Chemical Society*, **77**, 3433.

Reid, R. C., Prausnitz, J. M., and Poling, B. E. (1987). *The properties of gases and liquids*. McGraw Hill, New York.

Rowlinson, J. S., and Swinton, F. L. (1982). *Liquids and liquid mixtures* (3rd edn), pp.191–229. Butterworth, London.

Saad, H. A., and Gulari, E. (1984). *Berichte der Bunsen-Gesellschaft für physikalische Chemie*, **88**, 834.

Schilling, W., and Franck, E. U. (1988). *Berichte der Bunsen-Gesellschaft für physikalische Chemie*, **92**, 631.

Schmidt, W. J., and Reid, R. C. (1986). *Journal of Chemical and Engineering Data*, **31**, 204.

Schmidt, W. J., and Reid, R. C. (1988). *Chemical Engineering Communications*, **64**, 155.

Scott, R. L., and van Konynenburg, P. H. (1970). *Discussions of the Faraday Society*, **49**, 87.

Sengers, J. V. (1994). In *Supercritical fluids* (ed. E. Kiran and J. M. H. Levelt Sengers), pp. 231–71. Kluwer, Amsterdam.

Simmons, G. M., and Mason, D. M. (1972). *Chemical Engineering Science*, **27**, 89.

Span, R., and Wagner, W. (1996). *Journal of Physical and Chemical Reference Data*, **25**, 1509.

Streett, W. B. (1983). In *Chemical engineering at supercritical fluid conditions*, (ed. M. E. Paulaitis, J. M. L. Penninger, R. D. Gray, and P. Davidson), pp. 3–30. Ann Arbor, Michigan. Tiltscher, H., Wolf, H., and Schelchsorn, J. (1984). *Berichte der Bunsen-Gesellschaft für physikalische Chemie*, **88**, 897.

Tolman, R. C. (1979). *The principles of statistical mechanics*. Dover, New York.

Treybal, R. E. (1980). *Mass transfer operations*. McGraw Hill, New York.

Tsekhanskaya, Y. V., Iomtev, M. B., and Mushkina, E. V. (1964). *Russian Journal of Physical Chemistry*, **38**, 1173.

Turner, R. E., and Betts, D. S. (1974). *Introductory statistical mechanics*. Sussex University Press, Brighton.

Tyrrell, H. J. V., and Watkiss, P. J. (1979). *Diffusion in liquids*. Butterworths, London.

van Deemter, J. J., Zuiderweg, F. J., and Klinkenberg, A. (1956). *Chemical Engineering Science*, **5**, 271.

van Konynenburg, P. H., and Scott, R. L. (1980). *Philosophical Transactions of the Royal Society of London*, **A298**, 495.

van Wasen, U., and Schneider, G. M. (1980). *Journal of Physical Chemistry*, **84**, 229.

Vesovic, V., Wakeham, W. A., Olchowy, G. A., Sengers, J. V., Watson, J. T. R., and Millat, J. (1990). *Journal of Physical and Chemical Reference Data*, **19**, 763.

Walker, D. F. G. (1995), PhD Thesis, University of Leeds.

Watson, J. T. R., Basu, R. S., and Sengers, J. V. (1980). *Journal of Physical and Chemical Reference Data*, **9**, 1255.

Weiser, K. (1957). *Journal of Physical Chemistry*, **61**, 513.

Index

Printed in the United Kingdom
by Lightning Source UK Ltd.
102156UKS00001B/119